BITCH

A Revolutionary Guide to
Sex, Evolution and the Female Animal

LUCY COOKE

doubleday

TRANSWORLD PUBLISHERS
Penguin Random House, One Embassy Gardens,
8 Viaduct Gardens, London SW11 7BW
www.penguin.co.uk

Transworld is part of the Penguin Random House group of companies
whose addresses can be found at global.penguinrandomhouse.com

First published in Great Britain in 2022 by Doubleday
an imprint of Transworld Publishers

A CIP catalogue record for this book
is available from the British Library.

ISBNs 9780857524133 (hb)
9780857524126 (tpb)

Typeset in 11.5/15pt Sabon MT Pro by Jouve (UK), Milton Keynes.
Printed and bound in Great Britain by Clays Ltd, Elcograf S.p.A.

The authorized representative in the EEA is Penguin Random House Ireland,
Morrison Chambers, 32 Nassau Street, Dublin D02 YH68.

Penguin Random House is committed to a sustainable future
for our business, our readers and our planet. This book is made
from Forest Stewardship Council® certified paper.

To all the bitches in my life
Thank you for the love and inspiration

AUTHOR'S NOTE ON LANGUAGE

Language evolves rapidly, and there is currently much conversation about the conflation of sex and gender terms. It is critical to use these terms appropriately and not to confuse them. Most scientists agree that non-human animals do not have gender. In this book, the terms female and male refer to an animal's biological sex. I do engage in anthropomorphizing, to an extent. Sometimes this is because these were the historical terms used. For example, I may refer to an animal's genitalia as being 'masculinized' or a brain being 'feminized' as this was the original scientific description. Such gendered terms needn't and shouldn't be used to describe animals' sex characteristics and behaviours in scholarly realms today. I also use gendered terms such as 'mother' and 'father' to describe animals, because these are the terms used by the scientists in question and most of my audience will understand what or who I refer to with these terms – for instance, 'mother' may mean the egg-producing parent of an individual animal. At other times, I have used anthropomorphic terms such as femme fatale, queen, lesbian, sister, lady and bitch for storytelling purposes, and readers needn't choose to replicate these labels in their academic work. I recognize that this anthropomorphizing can, unintentionally, have gendered implications. This book intends to demonstrate that sex is wildly variable and that gendered ideas based on assumptions of binary sex are nonsense. It is my sincerest hope that this intent has been clearly communicated.

CONTENTS

Lucy Cooke, far left, with (from left to right) Mary Jane West-Eberhard, Sarah Blaffer Hrdy and Jeanne Altmann

INTRODUCTION

Studying zoology made me feel like a sad misfit. Not because I loved spiders, enjoyed cutting up dead things I'd found by the side of the road or would gladly root around in animal faeces for clues as to what their owner had eaten. All my fellow students shared the same curious kinks, so there was no shame there. No, the source of my disquiet was my sex. Being female meant just one thing: I was a loser.

'The female is exploited, and the fundamental evolutionary basis for the exploitation is the fact that eggs are larger than sperms,' wrote my college tutor Richard Dawkins in his bestselling evolutionary bible, *The Selfish Gene*.

According to zoological law, we egg-makers had been betrayed by our bulky gametes. By investing our genetic legacy in a few nutrient-rich ova, rather than millions of mobile sperm, our forebears had pulled the short straw in the primeval lottery of life. Now we were doomed to play second fiddle to the sperm-shooters for all eternity; a feminine footnote to the macho main event.

I was taught that this apparently trivial disparity in our sex cells laid cast-iron biological foundations for sexual inequality. 'It is possible to interpret all other differences between the sexes as stemming from this one basic difference,' Dawkins told us. 'Female exploitation begins here.'

Male animals led swashbuckling lives of thrusting agency. They fought one another over leadership or possession of females. They shagged around indiscriminately, propelled by a biological imperative

to spread their seed far and wide. And they were socially dominant; where males led, females meekly followed. A female's role
was as selfless mother, naturally; as such, maternal efforts were
deemed all alike: we had zero competitive edge. Sex was a duty
rather than a drive.

And as far as evolution was concerned it was males who drove
the bus of change. We females could hop on for a ride thanks to
shared DNA, as long as we promised to keep nice and quiet.

As an egg-making student of evolution, I couldn't see my reflection in this fifties sitcom of sex roles. Was I some kind of female
aberration?

The answer, thankfully, is no.

A sexist mythology has been baked into biology, and it distorts
the way we perceive female animals. In the natural world female
form and role varies wildly to encompass a fascinating spectrum of
anatomies and behaviours. Yes, the doting mother is among them,
but so is the jacana bird that abandons her eggs and leaves them to
a harem of cuckolded males to raise. Females can be faithful, but
only 7 per cent of species are sexually monogamous, which leaves a
lot of philandering females seeking sex with multiple partners. Not
all animal societies are dominated by males by any means; alpha
females have evolved across a variety of classes and their authority
ranges from benevolent (bonobos) to brutal (bees). Females can
compete with each other as viciously as males: topi antelope engage
in fierce battles with huge horns for access to the best males, and
meerkat matriarchs are the most murderous mammals on the planet,
killing their competitors' babies and suppressing their reproduction. Then there are the femme fatales: cannibalistic female spiders
that consume their lovers as post- or even pre-coital snacks and
'lesbian' lizards that have lost the need for males altogether and
reproduce solely by cloning.

In the last few decades there has been a revolution in our understanding of what it means to be female. This book is about that
revolution. In it, I will introduce you to a riotous cast of remarkable
female animals, and the scientists that study them, who together

have redefined not just the female of the species, but the very forces that shape evolution.

To understand how we arrived at this cockeyed view of the natural world, we have to head back in time to Victorian England to meet my scientific idol: Charles Darwin. Darwin's theory of evolution by natural selection explained how the rich variety of life is descended from a common ancestor. Organisms that are more adapted to their environment are more likely to survive and pass on the genes that aided their success. This process causes species to change and diverge over time. Often misquoted as 'survival of the fittest' – a term coined by the philosopher Herbert Spencer and only incorporated by Darwin under duress into the fifth edition of *On the Origin of Species* (1869) – the idea is as brilliant as it is simple and justly hailed as one of the greatest intellectual breakthroughs of all time.

As ingenious as it is, natural selection cannot account for everything we find in nature. Darwin's evolutionary theory had some gaping holes in it, caused by elaborate traits like the stag's antlers or the peacock's tail. Such extravagances offer no benefit to the general process of being, and could even be considered a hindrance to everyday life. As such they could not have been sculpted by the utilitarian force of natural selection. Darwin recognized this, and for a long time it tortured him. He realized there must be another evolutionary mechanism at play, with a very different agenda. That, Darwin eventually realized, was the quest for sex – and so he named it *sexual* selection.

To Darwin this novel evolutionary force explained these flamboyant traits – their only purpose must be to win or attract the opposite sex. To mark their non-essential nature Darwin christened such indulgences 'secondary sexual characteristics', to separate them from 'primary sexual characteristics', like reproductive organs and genitals, which are instead quite indispensable for perpetuating life.

Just over a decade after he presented natural selection to the world, Darwin published his second great theoretical masterpiece: *The Descent of Man, and Selection in Relation to Sex* (1871). This hefty follow-up tome outlined his new theory of sexual selection, which accounted for the profound differences he observed between the sexes. If natural selection is the struggle for survival, sexual selection is essentially the struggle for mates. And as far as Darwin was concerned, this competition was largely the domain of males.

'The males of almost all animals have stronger passions than the females. Hence it is the males that fight together and sedulously display their charms before the females,' Darwin explains. 'The female, on the other hand, with the rarest of exceptions, is less eager than the male . . . she generally "requires to be courted"; she is coy.'

Thus, in Darwin's eyes, sexual dimorphism also extended to the behaviour of each sex. These sex roles were as predictable as physical characteristics. Males take the evolutionary lead by duking it out with 'weapons' or 'charms' specially evolved in order to take 'possession' of the female. Competition is such that males will vary wildly in their reproductive success and this sexual selection drives the evolution of winning traits. Females have less call for variation; their role is one of submission to and transmission of these male characteristics. Darwin wasn't sure why this disparity existed, but he suspected it could be traced back to the sex cells and the female being energetically drained by her maternal investment.

In addition to male competition, Darwin knew that the mechanics of sexual selection required an element of female choice. This was trickier to explain because it gave the fairer sex an uncomfortably active role in shaping the male – something which would not go down well in Victorian England and, as we shall discover in chapter two, ultimately made Darwin's theory of sexual selection distinctly unpalatable to the scientific patriarchy. So Darwin was at great pains to downplay this female power by stating that it is somehow achieved in a 'comparatively passive' and unthreatening fashion by females 'standing by as spectators' to the masculine battle of bravado.

Darwin's branding of the sexes as active (male) and passive (female) could not have been more effective if it had been devised by a multi-million-dollar marketing company with an unlimited budget. It subscribes to the kind of tidy dichotomy – like right or wrong, black or white, friend or foe – so relished by the human brain for feeling intuitively correct.

But Darwin was probably not the originator of this convenient sexual classification. He likely borrowed it from Aristotle, the father of zoology. In the fourth century BC the ancient Greek philosopher wrote the first ever animal almanacs: *On the Generation of Animals* was his treatise on reproduction. Darwin had certainly read this seminal academic work, which perhaps explains why there is a distinct whiff of familiarity in Aristotle's partitioning of the sex roles.

'In those animals that have . . . two sexes . . . the male stands for effective and active . . . and the female . . . for the passive.'

The stereotypes of female passivity and male vigour are as old as zoology itself. Such an endurance test of time suggests they've 'felt right' to generations of scientists, but that doesn't mean they are. One thing science in every domain has taught us is that our intuitions often lead us astray. The main problem with this neat binary classification is: it's wrong.

Try explaining the need to be passive to a dominant female spotted hyena, and she'll laugh in your face, after she's bitten it off. Female animals are just as promiscuous, competitive, aggressive, dominant and dynamic as males. They have equal right to drive the bus of change. It's just that Darwin, along with the coterie of gentlemen zoologists that helped inform his argument, couldn't, or perhaps wouldn't, see them that way. The greatest single leap forward in all of biology – perhaps all of science – was made by a group of Victorian men, in a mid-nineteenth-century milieu, and it smuggled with it certain assumptions about the nature of gender and sex.

It's fair to say that if Darwin was a contestant on *Mastermind*, his specialist subject would not be the opposite sex. Here was a man that married his first cousin Emma, only after drawing up a list of nuptial pros and cons. This revealing romantic inventory, scribbled on the back of a letter to a friend, has, to Darwin's shame, been preserved, revealing his most intimate thoughts for all to judge in perpetuity.

In just two brief columns – 'Marry' and 'Not Marry' – Darwin thrashed out his inner connubial turmoil. His chief concerns were that he would miss out on the 'conversation of clever men at clubs' and might therefore succumb to 'fatness and idleness', or worse, 'banishment and degradation with indolent idle fool' (which is perhaps not the way Emma would have chosen to be described by her beloved fiancé). However, on the plus side, he would have 'someone to take care of the house' and a 'nice soft wife on a sofa' was 'better than a dog anyhow'. So Darwin bravely took the plunge.

One gets the feeling that, despite fathering ten children, Darwin was perhaps driven by cerebral rather than carnal urges. He may not have been terribly familiar with or even curious about the female sex. So the chances of him questioning evolution from the female perspective, as well as the male, were perhaps small, even before you consider the society into which he was born.

Even the most original and meticulous scientists are not immune to the influence of culture, and Darwin's androcentric reading of the sexes was no doubt shaped by the prevailing chauvinism of the era. Women in upper-class Victorian society had one main role in life: to marry, have children and perhaps assist with their husbands' interests and business. This was very much a supportive, domestic role since they were defined, physically and intellectually, as the 'weaker' sex. Women were, in all ways, subordinate to male authority, be that of fathers, husbands, brothers or even adult sons.

This social prejudice was conveniently substantiated by contemporary scientific thinking. The leading academic minds of the Victorian era considered the sexes to be radically different creatures – essentially polar opposites of one another. Females were believed to experience arrested development; they resembled the young of their species by being smaller, weaker and less colourful. Where male energy goes into growth, female energy is required to nourish eggs and carry young. Because of males' generally larger physique, they were considered to be more complex and variable than females, as well as superior in mental capacity. Females were considered to be all of average intelligence, but males varied wildly to include levels of genius unseen in the opposite sex. Essentially, males were considered to be more *evolved* than females.

These sentiments were all incorporated by Darwin into *The Descent of Man, and Selection in Relation to Sex*, which, as the title suggests, used sexual and natural selection to explain human evolution and the sex differences upheld by Victorian society.

'The chief distinction in the intellectual powers of the two sexes is shewn by man's attaining to a higher eminence, in whatever he takes up, than can woman – whether requiring deep thought, reason or imagination, or merely the use of the senses and hands,' explained Darwin. 'Thus man has ultimately become superior to woman.'

Darwin's theory of sexual selection was incubated in misogyny, so it is little wonder that the female animal came out deformed; as marginalized and misunderstood as a Victorian housewife. What is perhaps more surprising, and damaging, is how tough it has been to wash this sexist stain out of science, and how far it has bled.

Darwin's genius has not helped. Because of his godlike reputation, biologists who followed in his wake have suffered from a chronic case of confirmation bias. They looked for evidence in support of the passive female prototype, and saw only what they wanted to see. When faced with anomalies, like the licentious promiscuity of the female lioness that enthusiastically mates scores of times a day during oestrus with multiple males, they studiously looked the other way. Or worse, as you will discover in chapter three, experimental results that didn't conform were manipulated with a statistical sleight of hand to conjure sideways support for 'the correct' scientific model.

A central tenet of science is the parsimonious principle, also known as Ockham's razor, which teaches scientists to trust in the

evidence and choose the simplest explanation for it, as it will probably be the best. Darwin's strict sex roles have forced an abandonment
of this fundamental scientific process as researchers are compelled
to dream up ever more tortuous excuses to explain away female
behaviours that deviate from the standard stereotype.

Take the pinyon jay, *Gymnorhinus cyanocephalus*. These cobalt-
blue members of the crow family live in noisy flocks of fifty to five
hundred birds in the western states of North America. Highly
intelligent creatures with such active social lives are likely to have
some means of ordering their busy society – a dominance network –
otherwise there would be chaos. The ornithologists John Marzluff
and Russell Balda, who studied the jays for over twenty years and
published an authoritative book on them in the 1990s, were interested in decoding the pinyon jay's social hierarchy. So they went in
search of the 'alpha male'.

This took some ingenuity. It transpired that male pinyon jays are
committed pacifists and rarely ever fight. So, the enterprising ornithologists built feeding stations loaded with tasty treats like greasy
popcorn and mealworms to try to incite some kind of territorial
war. But still the jays refused to engage in battle. The researchers
were forced to base their scale of combat on some fairly subtle
cues, like sideways glances. If the dominant male gave the submissive male what amounted to a dirty look then the submissive would
leave the feeder. It wasn't exactly *Game of Thrones* stuff, but the
researchers sat and diligently recorded around two and a half thousand of these 'aggressive' encounters nevertheless.

When they came to run the statistics they were further confused.
Only fourteen of two hundred flock members qualified for a place
in the dominance network and there was no linear hierarchy.
Males reversed their dominance and subordinates 'aggressed' their
superiors. Despite the puzzling results and general lack of macho
hostility, the scientists still felt confident in declaring, 'There is little
doubt that adult males are in aggressive control.'

The curious thing is, the researchers had seen jays behaving with
significantly more antagonism than a few annoyed looks. They

documented birds in dramatic airborne battles where duelling pairs became locked in combat mid-air and 'flap vigorously as they fall to the ground' where they 'peck at each other with forceful stabs'. These encounters were 'the most aggressive behaviour observed during the year', but they were not included in any dominance network as the perpetrators weren't male. They were all female. The authors concluded that this 'testy' feminine behaviour must be hormonally driven. They proposed that a spring hormone surge had given these female jays 'the avian equivalent of PMS which we call PBS (pre-breeding syndrome)'!

There is no such thing as avian PBS. If Marzluff and Balda had had their minds open to the female birds' aggressive behaviour and used Ockham's razor to shave the fluff from their conjecture, they would have got close to figuring out the pinyon jay's complex social system. The clues that females are in fact highly competitive and play an instrumental role in the jay's hierarchy are all there in their meticulously recorded data, but they were blind to them. Instead they pushed forward dogmatically in search of 'the crowning of a new king', a coronation of their conviction which, of course, never happened.

There is no conspiracy here, just blinkered science. Marzluff and Balda illustrate how good scientists can suffer bad biases. The ornithological duo were faced with confounding novel behaviour, which they interpreted within a bogus framework. They are by no means alone in their honest error. Science, it transpires, is soaked in accidental sexism.

It hasn't helped that the academic establishment was, and in many areas still is, dominated by men who naturally view the animal kingdom from their standpoint; the questions asked to inspire research thus originated from a male perspective. Many simply weren't curious about females. Males were the main event and became the model organism – the default from which the female deviated, the standard by which the species was judged. Female animals, with their 'messy hormones', were the outliers, distracting tangents to the leading narrative, and didn't warrant the same level of scientific scrutiny. Their bodies and behaviours were left unexamined. The resulting data gap then becomes a self-fulfilling prophecy. Females are seen as the invariant and inert sidekicks to male endeavour, because there's no data to sell them as anything otherwise.

The most dangerous thing about sexist bias is its boomerang nature. What started as chauvinist Victorian culture was incubated by a century of science and then spat back into society as political weaponry, rubber-stamped by Darwin. It gave a handful of, notably male, devotees of the new science of evolutionary psychology the ideological authority to claim that a host of grim male behaviours – from rape to compulsive skirt chasing to male supremacy – were 'only natural' for humans, because Darwin said so. They told women they had dysfunctional orgasms, that they could never break through the glass ceiling thanks to an innate lack of ambition, and should stick to mothering.

This turn of the century evolutionary psychobabble was gobbled up by a new breed of men's magazines, that shunted this sexist 'science' into the mainstream. In bestselling books and high-profile columns in the popular press, journalists like Robert Wright crowed that feminism was doomed because it refuses to acknowledge these scientific truths. From his ideological pedestal Wright penned imperious articles with titles like 'Feminists, Meet Mr Darwin' and awarded his critics 'a C in Evolutionary Biology 101', claiming that 'not a single well-known feminist has learned enough about modern Darwinism to pass judgement on it'.

But they had. The second wave of feminism had opened once-closed laboratory doors and women were walking the halls of top universities and studying Darwin for themselves. They were heading into the field and observing female animals with the same curiosity as male animals. They discovered sexually precocious female monkeys and, instead of ignoring them like their male predecessors had, they questioned why they might be behaving in this way. They developed standardized techniques for measuring behaviour that forced equal attention on *both* sexes. They harnessed new technologies to spy on female birds and reveal that far from being victims of male sexual dominance, they were in fact running the show. And they repeated experiments that empirically underpinned Darwin's sexual stereotypes and discovered the results had been skewed.

It takes courage to challenge Darwin. He's more than just an iconic intellect; he's a national treasure in the UK. As one veteran professor pointed out to me, disagreeing with Darwin is tantamount to academic heresy and has led to a distinct conservatism in our homegrown evolutionary science. It is perhaps for this reason that the first seeds of rebellion came from the other side of the Atlantic, and a sprinkling of American scientists who ventured to originate alternative narratives about evolution, gender and sexuality.

You will meet these intellectual warriors in the pages of this book. I met some of them over lunch at a walnut farm in California where we discussed Darwin, orgasms and vultures, amongst other things. Sarah Blaffer Hrdy, Jeanne Altmann, Mary Jane West-Eberhard and Patricia Gowaty are the rabble-rousing matriarchs of modern Darwinism who dared to fight the scientific phallocracy with data and logic. They call themselves 'The Broads' and have met privately at Hrdy's home every year for the last thirty years to chew the evolutionary fat. I lucked out by landing an invitation to their annual cerebral jamboree. Although now semi-retired, these trailblazing professors still gather to support one another, discuss fresh ideas and generally keep the course of evolutionary biology evolving on an even path. They are feminists, yes, but they are clear that means they believe in the equal representation of both sexes, not the undeserved dominion of one.

Their science has enabled a new wave of biologists to look at the female of the species as fascinating in her own right; by examining female bodies and behaviour and asking questions about how selection works from the perspective of a daughter, sister, mother and competitor. These scientists have been willing to look beyond cultural norms and entertain unorthodox ideas about the fluidity of sex roles, overthrowing the machismo – inadvertent or otherwise – of evolutionary biology. Many are female, but, as you will discover, this scientific mutiny is not a women-only space – all sexes and genders are playing a part. You will meet many male scientists in the pages of this book. The pioneering work of Frans de Waal, William Eberhard and David Crews, to name just a few, proves that you don't need to identify as female to be a feminist scientist. And fresh perspectives from the LGBTQ scientific community have been crucial in challenging zoology's heteronormative myopia and the binary dogma. Biologists like Anne Fausto-Sterling and Joan Roughgarden, amongst others, have drawn attention to the stunning variety of sexual expression in the animal kingdom, and diversity's fundamental role in driving evolution forward.

The result is not just a more fabulously rich and life-like portrait of the female animal, but also a wealth of surprising new insights into the tangled mechanics of evolution. These are exciting times for evolutionary biologists: sexual selection is in the throes of a major paradigm shift. Empirical revelations are turning accepted facts on their head and conceptual changes are sending long-held assumptions out of the window. Darwin wasn't all wrong on this score, by any means. Male competition and female choice *do* drive sexual selection, but they are just part of the evolutionary picture. Darwin was viewing the natural world through a Victorian pinhole camera. Understanding the female sex is giving us the widescreen version of life on earth, in full technicolour glory, and the story is all the more fascinating for it.

In *Bitch* I go on a global adventure to meet the animals and scientists that are helping to rewrite an outdated patriarchal view of evolution and redefine the female of the species.

I travel to the island of Madagascar to discover how female lemurs, our most distant primate cousins, came to dominate males physically and politically. In the snowy mountains of California I discover how a robot female sage grouse exploded Darwin's myth of the passive female. On the island of Hawaii I meet loved-up, long-term female albatross couples that have defied traditional sex roles and shacked up together to raise their chicks. And cruising along the Washington coast, I find kinship with a matriarchal killer whale – the wise old leader of her hunting community, and one of only five known species, including humans, in which females go through menopause.

By exploring emerging tales from the fringes of femaleness I hope to paint a fresh, diverse portrait of the female animal, and to try to understand what, if anything, these revelations can inform us about our own species.

Since the time of Aesop, humans have looked to animals as illustrations and models of human behaviour. Many believe, somewhat misguidedly, that nature teaches human societies what is good and correct – the naturalistic fallacy. But survival is an unsentimental sport and animal behaviour encompasses female narratives that range from the fabulously empowered to the terrifyingly oppressed. Scientific discoveries about female animals can be used to fuel battles on both sides of the feminist fence; wielding animals as ideological weapons is a dangerous game. But understanding what it means to be a female animal can help counter lazy arguments and tired androcentric stereotypes; it can challenge our assumptions about what is natural, normal and even possible. If womanhood is going to be defined by one thing, rather than strict, outdated rules and expectations, it is its dynamic and varied nature.

The bitches in *Bitch* will demonstrate how being female is about being a fighter for survival and not just a passive sidekick. Darwin's theory of sexual selection drove a wedge between the sexes by focusing on our differences; but these differences are greater culturally than they are biologically. Animal characteristics – be they physical or behavioural – are both varied and plastic. They can bend according to a selection's whim, which makes sex traits fluid and malleable. Rather than predicting a female's qualities through the crystal ball of her sex, the environment, time and chance all play a significant role in shaping their form. As we shall discover in the first chapter, females and males are, in fact, far more alike than they are different. So much so, it can sometimes be hard knowing where to draw the line.

CHAPTER ONE

The anarchy of sex: what is a female?

Let's start by heading underground to meet a highly secretive female: enemy number one of the landscape gardener and greedy consumer of worms. I'm talking about the mole, *Talpa europaea*.

Most of you will be familiar with the mole's handiwork, if not the beast itself. Their conical piles of freshly turned dirt can disrupt a smoothly manicured lawn like a chronic case of acne – the ultimate pain in the grass.

Back in the 1970s my father would be driven to distraction by molehills invading his precious turf. Much to my dismay, he'd set barbaric-looking metal traps to catch their creators. Once a mole was ensnared, I would insist he hand their lifeless bodies over to me so I could stroke their oh-so-velvety silver-black fur and marvel at their strangeness – their minute beady eyes (which despite popular mythology are poorly sighted but not totally blind) and comically oversized pink front paws – before giving them a proper burial. Back into the earth, where they belong.

The female mole is indeed a wondrous creature. A solo operator who makes her living by hunting worms using a network of tunnels that act as her own form of animal trap. When a worm pushes through her subway ceiling, she quickly sniffs it out using a long pink snout that can actually smell in stereo – each nostril acts independently, allowing her brain to accurately compute the direction of dinner in the pitch black. Her quarry, once caught, isn't killed immediately; instead, the mole paralyses it with her venomous saliva so it can be stored alive in a specially constructed larder without turning to rot. As many as four hundred and seventy wrigglers have been recorded in one lucky mole's pantry, which is helpful as she needs to consume over half her body weight in worms a day.

Life underground is tough. Burrowing earth is exhausting work and there's comparatively little oxygen to breathe. To survive this hostile environment evolution has equipped the mole with some cunning specializations. Her blood sports a modified form of haemoglobin that increases her affinity for oxygen and tolerance of toxic waste gases. And she sports an extra 'thumb'. Just like in the panda, a bone from her wrist has shot off on its own evolutionary path and formed a useful new digit for shifting extra earth. But perhaps most impressive of all are the female mole's balls.

The mole sow's gonads are described as 'ovotestes'. These internal reproductive organs consist of ovarian tissue at one end and testicular tissue at the other. The ovary side produces eggs and expands during the short breeding season. But, once the job of reproduction is done, this egg-making tissue shrinks and the testicular tissue expands until it is actually larger than the ovarian.

The female mole's testicular tissue is full of Leydig cells that make testosterone, but not sperm. This sex steroid hormone is commonly associated with males: beefing up muscles and fuelling aggression. It does both in the female mole, giving her the evolutionary edge underground: extra digging power and added hostility for defending her pups and worm larder.

It also gives her genitalia that are indistinguishable from the male's: an enlarged clitoris variously described as a 'phallus' or 'penile clitoris' and a vagina that seals up outside of breeding.

The female mole forces us to confront age-old assumptions about what distinguishes the sexes. For the majority of the year, on a genital, gonadal and hormonal level, the mole sow could easily be mistaken for a boar. So, how do we know she's a female?

This is a book about non-human animals, so it is important to begin by separating sex and gender. Most biologists agree that animals don't have gender. This social, psychological and cultural construct is considered the preserve of humans. When biologists talk about females they are referring only to their sex, but what does that mean?

In the beginning, reproduction was simple. The earliest life forms simply split, fused, budded bits off or cloned themselves in order to multiply. Then along came sex, which complicated matters somewhat. Now individuals needed to combine sex cells – gametes – in order to proliferate. Across the animal kingdom these come in just two sizes: big and small. This basic gametal dichotomy provides the standard biological definition of sex: females produce large, nutrient-rich eggs and males make small mobile sperm.

So far, so binary. Or is it?

Well, no. Sex is a complicated business. As you will discover in this first chapter, the ancient network of genes and sex hormones that interact to determine and differentiate the sexes have the ability to create a mixture of gametes, gonads, genitals, bodies and behaviour that disregards binary expectations. All of which makes marshalling sex into two neat deterministic buckets far from straightforward.

Starting at the most superficial level, many would consider genitals an easy indicator of sex. But the female mole's 'phallus' blows that notion clean out of the water. She's no freak. Dozens of female animals, from tiny cave-dwelling barklice* to giant African elephants,

---

* Two distinct genera of barklice, *Neotrogla* in South America and *Afrotrogla* in southern Africa, have evolved fully erectable 'penises' in females and 'vaginas' in males. Females of these cave-dwelling insects are the more promiscuous and aggressive sex. They are about the size of a flea and use their tiny, spiny penis to anchor on to the male during sex. This can last for 40–70 hours, during which time a capsule of sperm travels from the male to the female. Given the geographical distance between these two barklice populations, it would appear that the female's sexual tackle evolved on two independent occasions and not from a single shared ancestor.

sport ambiguous sexual anatomy that's commonly described in phallic terms.

The first time I saw a female spider monkey in the Amazon I assumed it was a male because of its dangling sexual appendage, the ostentatious size of which seemed to me frankly hazardous as it cavorted about the canopy. The primatologists I was with politely corrected me. Male spider monkeys are the sex with no apparent penis, since they keep theirs tucked away inside. Females on the other hand have a very obvious pendulous clitoris, known in biological circles as a 'pseudo-penis'. Such androcentric terminology grates somewhat, especially when you consider that the female spider monkey's 'fake' phallus is in fact longer than the male's 'real' phallus.

The strangest example is perhaps the fossa. Madagascar's greatest predator is the largest member of the mongoose family and looks a bit like a puma with a shrunken head. Its scientific name, *Cryptoprocta ferox*, translates as 'ferocious, hidden anus'. That taxonomists chose to highlight the fossa's anus as cryptic is somewhat unconventional, when it's the rest of her privates that are so mysterious.

When a female fossa is born she has a small clitoris and vulva, as might be expected. Then, at around seven months of age, something odd starts to happen. The fossa's clitoris enlarges, grows an internal bone and acquires spines to become a facsimile of the male's penis. It even exudes yellow liquid on its underside, like an adult male's. The female fossa sports her penile clitoris for a year or two, until she becomes reproductively active, when it magically disappears. Authors of a scientific paper on fossa genitalia postulated that this might protect the adolescent female from the unwanted attention of sexually pushy males or aggressively territorial female fossas.

The female fossa's transitory flirtation with a lookalike penis might, of course, serve no function at all. Not all traits do. Much like the redundant human appendix, it could simply be a relic from the fossa's evolutionary past that was sufficiently benign to avoid being selected against. Or be a side effect of another trait that evolution has selected for. Deciphering the ultimate evolutionary cause of a novel characteristic is a speculative sport. But decades of study into a close relative of the fossa has provided valuable evidence for the mechanics underlying such 'masculinized' genitalia. These insights have challenged a long-standing scientific prejudice concerning the 'passive' nature of female sexual development and gendered stereotypes of the hormones involved.

The genitals of the spotted hyena, *Crocuta crocuta*, have been causing a stir since the time of Aristotle. Ancient naturalists believed the hyena to be a hermaphrodite on account of the female's pudenda, which are the most sexually ambiguous of any known mammal's. Not only does the female spotted hyena have an eight-inch clitoris that's shaped and positioned exactly like the male's penis but she also gets erections. Both female and male spotted hyenas display and inspect one another's sexual tumescence during 'greeting ceremonies'. Crowning all this female virility is what appears to be a prominent pair of furry testicles.

This scrotum is in fact false: the hyena's labia have fused and filled with fatty tissue and merely resemble male gonads. This means that the female spotted hyena is the only mammal with no external vaginal opening at all. Instead she must urinate, copulate and even give birth through her curious multi-tasking clitoris – hence the antiquated hermaphrodite rumour. In more recent years, scientists have noted that males and females are so similar that they can be differentiated only by 'palpation of the scrotum' – something of a last resort, one assumes, when sexing an animal famous for its bone-crunching bite.

The female spotted hyena's sexual transgression doesn't stop at her genitals. Scientists have also been fascinated by her similarly 'masculinized' body and behaviour. Females can be up to 10 per cent heavier than males in the wild (20 per cent in captivity). This is unusual, as amongst mammals males are generally larger in size.* In the rest of the animal kingdom, and thus the majority of animals, sexual size dimorphism is however generally the reverse. Fatter, more fecund females produce more eggs, so amongst most invertebrates and many fish, amphibians and reptiles it is the females that often outsize the males.†

Female spotted hyenas are also more aggressive than males. These highly intelligent, social carnivores live in matrilineal clans of up to eighty individuals governed by an alpha female. Males tend to be the sex that disperses from the natal matriline and, as such, the lowest rung of hyena society: submissive outcasts begging for acceptance, food and sex. Females are considered dominant in most situations, engage in rough play and vigorous scent-marking as well as leading the territorial defence – all behaviours more commonly associated with the opposite sex.

The radical gender-bending life of the female spotted hyena was originally assumed to be the result of an excess of testosterone circulating in her blood. Androgens, the group of sex steroid

---

* There are other female mammals that also buck the size trend. The most extreme mammalian case is a South American bat, *Ametrida centurio*, whose males are so much smaller they were originally classified as a separate species. This 'reversed' size dimorphism may be linked with an aerial lifestyle as it is also common in birds – the rationale being that competing males need to be agile rather than strong, and thus evolve to be smaller in size than females. At the other end of the scale, many species of baleen whale have bigger females than males, including the blue whale. One female specimen taken off the coast of the island of South Georgia was almost 30 metres long and weighed in at 173 tonnes – three times the length of a double-decker bus and over thirteen times its weight. Which means that the largest animal to have ever lived was, in fact, female.

† The deep-sea angler fish *Ceratias holboelli* has taken this to an extreme. Males may be more than sixty times shorter and half a million times lighter – essentially little more than swimming sacs of sperm. Once the male has sniffed out a female from her leaking pheromones in the pitch-black depths, he will latch on to her with his mouth, physically fusing with her body for the rest of his life – the evolutionary embodiment of a clingy sexual freeloader. The female is thereafter in control of the male's entire existence, including when he ejaculates his sperm. The Danish fisherman who discovered this intimate set-up in 1925 declared that 'so perfect and complete is the union of husband and wife that one may almost be sure that their genital glands ripen simultaneously'. And they say romance is dead.

hormones that includes testosterone, have been unambiguously branded as male: andro meaning 'man' and gen a 'thing that produces or causes'. So the obvious assumption was that these big, belligerent female hyenas, much like the mole we met earlier, must be swilling with the stuff. But much to everyone's surprise, the circulating levels of testosterone in adult female spotted hyenas do not rival those in males.

So where was all this virilization coming from? The bitches' pseudo-penises pointed to a different timing for testosterone's influence, namely during fetal development.

The standard paradigm for sexual differentiation was developed in the 1940s and '50s by a French embryologist named Alfred Jost, following a series of pioneering, if barbarous, experiments on rabbit fetuses at various stages of development in their mother's womb.

Mammal embryos, whether they're female or male, all start off with a unisex kit of parts: an assortment of ducts, tubes and proto-gonadal tissue with the potential to develop into either ovaries or testes. The developing fetus is thus considered sexually 'neutral' until this primordial sexual medley starts its journey down the ovarian or testicular path.

Jost's experiments on developing rabbits didn't figure out what triggers the initial differentiation (more on that later), but he did establish that testosterone plays a primary role in driving the fetal gonad towards becoming testes and the subsequent development of male genitalia.

Jost discovered that if he removed the male embryonic gonads early in development, the fetus failed to grow a penis and scrotum and developed a vagina and clitoris instead. Removing the developing ovaries of a fetal female, on the other hand, did not obviously impact her sexual development. Oviducts, uterus, cervix and vagina all developed in an apparently automatic fashion without the need of her embryonic ovaries or their hormones to direct them. In contrast, just 'a crystal of androgen could counteract the absence of testicles' and ensure the development of male sexual characteristics, heralding this sex steroid to be the dynamic elixir of maleness.

By a process of elimination over dozens of experiments Jost established that high concentrations of testosterone in the male fetus, produced by the developing testicular cells, actively pushed an embryo down the path of male sexual development. In contrast, the creation of a female was seen as a passive process – the 'default' result of an absence of gonadal testosterone.

Jost's theory slotted in nicely with the widespread notion, popularized by Darwin, that females were generally passive and males active. The theory was embellished by others and labelled the Organizational Concept – the universally accepted model for sexual differentiation not just of bodies, but behaviour too. It placed male gonads and androgens in the starring role – the saviours of the sexual paradigm and chief architects of all things male.

Testes and their testosterone-pumping powers became the engine driving the demarcation of not just the embryonic gonads and genitals, but also the fetal neuroendocrine system and developing brain. This then programmed sexual differences in bodies and behaviour that could be activated by sex steroid hormones in later life. Thus testosterone became the executive director of sexual dimorphism; responsible for characteristics ranging from the hefty horns on the stag to the bull elephant's raging musth and the male walrus's fearsome size and temper.

Jost's findings revolutionized the ongoing debates in endocrinology on the hormonal origins of masculinity and femininity. At a conference in 1969, Jost explained: 'Becoming a male is a prolonged, uneasy and risky adventure; it is a kind of struggle against inherent trends toward femaleness.'

The masculine journey was seen as a heroic quest worthy of investigation. In contrast, the now-famous French embryologist referred to females simply as the 'neutral' or 'anhormonal' sex type. Ovaries and oestrogen were considered irrelevant to our story: inert and insignificant. Our sexual development was unreactive and scientifically trivial. Females basically 'just happened' because we lacked the embryonic balls to be male.

This prejudice has been remarkably enduring and damaging. The legacy of the Organizational Concept is an understudied female system and an unyielding binary view of sexual differentiation, as promoted by the all-powerful developmental male-wash of testosterone. But then along came the spotted hyena with her big phallic clitoris to suggest there's trouble in the paradigm.

Testosterone is indeed a potent hormone. If delivered at the right time, it has the power to reverse the gonadal sex of female fish, amphibians and reptiles. In mammals, it can't force a full sexual U-turn but marinating a female fetus in androgens radically alters the formation of her genitals. Experiments in the 1980s created female rhesus monkeys with a penis and scrotum 'indistinguishable from that of males' by exposing them to testosterone at key stages during their gestation.

Sure enough, when tested, female spotted hyenas revealed rocketing levels of testosterone during their pregnancy. But, with no testes in sight, what could be the source of this 'male' hormone, and how does the developing female fetus manage to survive its omnipotence yet still develop a functioning reproductive system?

The answers lie in how testosterone is synthesized. All of the sex hormones – oestrogen, progesterone and testosterone – originate as cholesterol. This steroid is converted by the action of enzymes to progesterone, a hormone commonly associated with pregnancy and the precursor to androgens, which are, in turn, the precursors to oestrogens. These 'male' and 'female' sex hormones can convert from one to the other and are present in both sexes.

'There's no such thing as a "male" hormone or a "female" hormone. It's a common misconception. We all have the same hormones,' Christine Drea revealed to me over Skype. 'All that differs between males and females are the relative amounts of enzymes that convert the sex steroids from one to another and the distribution and sensitivity of hormone receptors.'

Drea is a professor at Duke University and knows more than

most about the hormonal politics of female sexual differentiation. She's devoted her career to studying a suite of so-called 'masculinized' females, including the spotted hyena along with meerkats and ring-tailed lemurs.

Drea is part of the team that established the source of the pregnant hyena's testosterone. It comes from a lesser-known androgen called androstenedione, or A4, that's actually produced by the pregnant female's ovaries. This form of androgen is known as a precursor hormone, as it converts to either testosterone or oestrogen following the action of enzymes in the placenta.

In most pregnant mammals carrying daughters, A4 is preferentially transformed into oestrogen, but in the spotted hyena it transforms to testosterone instead. This 'male' hormone then exerts its influence on the developing genitals and brain of the female fetus, transforming both her pudenda and her post-natal behaviour.

Historically, A4 aroused little interest as a sex hormone; it was dismissed as 'inactive' for not binding to known androgen receptors. But receptors are now being located that suggest it does have direct action and, more crucially, its effects may differ depending on the sex of the fetus.

'There's a growing body of literature suggesting that hormones can have sexually differentiated effects in different animals. It's all about amount, duration and timing,' Drea articulated.

Drea's work clearly demonstrates that making a female is far from a 'passive' process, and one in which androgens can play an active role. 'Testosterone is not a "male" hormone. It is just a hormone that is more obviously expressed in males than females,' she reiterated.

It's clear to Drea that the female hyena's sexual development must also be under dynamic genetic control to resist the overpowering effects of an excess of androgens and still create a functional, if eccentric, reproductive system. But how is still much of a mystery. The functional genetic steps of how to actually make female reproductive organs are still poorly understood when compared to the male.

This bias in our understanding comes from Jost's famous but flawed theory of sex differentiation, which only ever explained how you differentiate a male and never questioned how the female is created. The idea that any development process could be 'passive' is clearly quite ludicrous – ovaries require just as much active assembly as do testes. Yet for fifty years the 'default' female system went unstudied.

'Sexual differentiation isn't about describing how you get females and males. It's only about describing how you get males. For decades people were happy not to have an explanation of how you get the female form and just saying, "Well, it's passive," ' Drea asserted.

A foundational publication on mammalian sexual development from 2007 referred to the development of the ovary as 'Terra Incognita'. The prevailing view that ovarian development is the 'default' state had, it claimed, led to 'a widespread understanding that no active genetic steps need to be taken to specify or create an ovary or female genitalia'. Which, the authors wryly note, is 'a rather amazing situation given the importance of this organ for proper female development and reproduction'.

Things are improving. The unknown land of ovarian development has now been partially explored, but its genetic map is far emptier than the one that exists for testes. The chauvinistic hangover of the Organizational Concept has focused the genetic quest for sexual determination firmly on the male; at its core was the hunt to find the elusive testis-determining factor, the genetic trigger that instructs those neutral fetal gonadal cells to rouse themselves out of their sexually indifferent slumber and transform into testes (and start pumping out testosterone).

The genetic recipe that actually determines the sexes is, however, positively byzantine in nature and features an ancient cast of surprisingly androgynous genes.

CHAOTIC CHROMOSOMES

You might think the ultimate answer to what makes a female animal is a pair of XX chromosomes. We're all taught in school, after all, how this anomalous pair of sex chromosomes defines the sexes, with males being XY and females XX. But sex is never that simple.

The XY sex-determination system is best known because it occurs in mammals, along with some other vertebrates and insects. In this system females have two copies of the same sex chromosome (XX), whereas males have two kinds of sex chromosomes (XY). The first misconception is that the letters X and Y refer to the shape of the chromosomes. All chromosomes are sausage-shaped and their resemblance, when paired, is entirely coincidental.

The very first X chromosome was discovered in 1891 by Hermann Henking, a young German zoologist who noticed something curious while inspecting the testicles of a fire wasp (which, to add further confusion, is a bug not a wasp). Chromosomes inhabit cells as matching pairs, but Henking noticed that in all the specimens he studied there was one chromosome that didn't appear to have a matching partner and remained aloof. He named it X – the mathematical symbol for unsolved – after its mysterious nature. Henking made no association between this now iconic, yet enigmatic, strand of DNA and sex determination, which is a shame as it could have made him quite famous. Instead, a year later, he abandoned his cytology studies and moved on to a career in fisheries, which was more financially rewarding but offered significantly less opportunity for scientific fame.

The Y chromosome was eventually discovered lurking in the reproductive organs of a mealworm some fourteen years later, in 1905, by the American Nettie Stevens – a pioneering female geneticist. Stevens recognized its key role in sex determination, but also received little fame for her epic breakthrough. The same chromosome was also discovered, more or less simultaneously, by a male scientist called Edmund Wilson, who sucked up most of the fame. It was eventually named Y to continue the alphabet system that Henking had started. But thanks to its peculiar stunted size, it also resembles the letter that provides its name when paired with the longer X.

Compared to the X, the Y is essentially a runt of a chromosome: stunted and with significantly less genetic material. When it comes to chromosomes, however, it's not size that matters, it's what you code with it. And the Y is indeed home to a very significant sex-determining gene called SRY (standing for Sex-determining Region of the Y).

In the 1980s Peter Goodfellow's lab in London finally unmasked this unassuming piece of genetic code as the elusive testis-determining factor in humans. His team discovered that the switching on of SRY proved to be the crucial first genetic step in triggering the neutral fetal gonad sex cells to develop into testes and start pumping out testosterone. In its absence, the unisex primordial kit matures at a more leisurely pace into embryonic ovaries.

This time there was much fanfare. The master switch for mammalian sex determination had finally been revealed and the 'essence of maleness' located. SRY was the missing trigger for the cascade of genes that code for testes development – the male sex-determining pathway.

I spoke to Jennifer Marshall Graves, the distinguished Australian professor of evolutionary genetics who was part of the international cohort of scientists hunting for this crucial male sex-determining gene. Her work on marsupial chromosomes prompted the search to switch direction to a fresh section of the Y, where the SRY gene was eventually located. Graves explained why their triumph at solving the puzzle of sex was, in fact, short-lived.

'We thought it was going to be the Holy Grail,' she confessed over Zoom from her home in Melbourne. 'When my student found the SRY gene we thought it would all be really simple. A kind of switch. But sex determination turns out to be much more complicated than we thought.'

The way sex is taught, you'd be forgiven for assuming the genes for creating testes inhabit the Y and the genes for ovaries reside on the X. That would be helpful. But evolution has done nothing to make the work of geneticists easy.

The entire process of sex organ determination involves an orchestra of around sixty genes working in concert. These sex-determining genes don't all exist on the sex chromosomes, let alone sit in a disciplined and gendered fashion on either the X or the Y. They are, in reality, scattered haphazardly throughout the genome.

SRY is like their conductor. If this crucial testes-determining trigger is present it instructs these sex-determining genes to start playing in the key of T for testes. If SRY is absent they'll play in the key of O for ovary. For a long time geneticists assumed these must be two entirely separate linear pathways, one for males (triggered by SRY's presence) and the other for females (triggered by SRY's absence). But the idea that evolution would produce such a tidy binary solution for sex proved to be woefully naive.

This is where sex becomes fabulously complicated. Aside from SRY, this orchestra of sixty sex-determining genes is basically the same in males and females. These genes have the ability to create either ovaries or testes, but exactly which gonad they actually produce depends on a complex network of inter-gene negotiation.

This kind of blew my mind. But Graves patiently spelt it out. 'A lot of these genes are not a "testes" gene or an "ovary" gene. They're kind of "both" genes and it depends on how many there are and which way they're driving the biochemical reactions. We're finding out all the time that some of these genes have more than one function at more than one stage.'

What's more, the two pathways – to either testes or ovaries – are neither linear nor separate. They're enmeshed. For example, some genes along the male path are needed to promote the development of the gonad in the direction of testes, whilst others are required to suppress the gonad heading in the direction of ovaries.

'It is overly simplistic to say that there's a single pathway that makes a testis, because there's also one that doesn't make an ovary at the same time. It's a whole contradictory mess of reactions, because there are so many genes that are intermediate – inhibiting one pathway and strengthening the other. So, these two sex "pathways" are intimately linked,' Graves explained.

In an effort to clarify this complexity, Graves sent me an animation of a crazy machine with dozens of interconnected ratchets and cogs all whirring around with little blue balls pinging in between them, and occasionally being squashed and recreated. The passage of the blue balls through this jumbled mess is her idea of how these purportedly neat binary sex-determination pathways really work.

This interconnected chaos of androgynous genes explains the plasticity of sex. Subtle tweaks in the expression of any of the interwoven cogs will produce novel variations – the grit that drives evolution forward and allows animals to adapt and exploit challenging new environments.

The female mole we met at the start of this chapter provides a handy illustration. A global consortium of scientists recently sequenced the entire genome of the Iberian mole, *Talpa occidentalis*. They compared the code with other mammals and found no differences in the protein products of the genes involved in sex determination. They did however discover mutations that altered the *regulation* of two of the sex-determination genes. These enabled a gene that's vital for developing testes to remain switched on in the female, as opposed to being inhibited. This accounts for the swollen section of testicular tissue in the sow's ovaries. In addition, another gene that codes for an enzyme involved in the production of androgens had two extra copies, increasing the female mole's testosterone output and allowing her to exploit the benefits of 'adaptive intersexuality'.

There is further variation still. SRY, the genetic trigger for this orchestra of sixty sex-determination genes, is not the universal master switch for sex across the animal kingdom, or even amongst all mammals for that matter.

Enter the platypus. This egg-laying mammal from Australia

specializes in being contrary, and its sex chromosomes are no different. Graves was part of the team that discovered how the platypus has five pairs of sex chromosomes. Females are XXXXXXXXXX and males XXXXXYYYYY. Despite this extravagance of Y chromosomes there is no sign of the SRY master switch on any of them.

'It was a shock,' Graves recalled.

The platypus is an ancient mammal whose group, the monotremes, diverged from humans some 166 million years ago. Its quirky sex chromosomes provided Graves with valuable insights into the evolution of sex chromosomes, and the shaky future of the Y.

The orchestra of sex-determining genes in the platypus, it transpires, is basically the same as it is in other mammals. Graves has discovered that these sixty or so genes are, in fact, remarkably conserved across all vertebrates. Birds, reptiles, amphibians and fish all have more or less the same set of genes as mammals for creating a testis or an ovary. What differs, however, is the master switch that kicks off the pathways. In the platypus this turned out to be one of the genes that's in the orchestra and has stepped up to the front to trigger the whole sex-determination process.

'SRY is just one way of kicking off the pathway, but you can do it by really almost any one of these sex-determining genes,' Graves explained, blowing my mind just a little bit more. 'That is the weirdest thing about sex. There are so many ways of doing it and they look to be quite different, but they're actually not. They all have to do with this pathway of sixty genes. So, the pathways are similar. But it's a completely different trigger.'

The platypus genome also revealed something else to Graves: the Y chromosome is losing genetic material. This runt of a chromosome is actually shrinking. Graves looked at how the platypus Y was different to the human Y and calculated how much genetic material had been lost in the time since our species diverged. This enabled her to estimate how long it might be before the human Y chromosome disappeared completely.

'It turned out the human Y was losing about ten genes per

million years, and there's only forty-five genes left. So it doesn't take Einstein to figure out at that rate we're going to lose the entire Y chromosome in four and a half million years.'

Certain high-profile geneticists, notably men, found the news that their 'male' chromosome was on a withering path to extinction rather hard to swallow.

'I thought it was hilarious. But David Page [eminent professor of genetics at MIT who contests Graves's prediction] did not think it was hilarious. He was apparently attacked by feminists saying, "Hey, you're all washed up!" To this day there is still that sort of crinkly animosity about the whole idea. And his desperate attempt to save the Y chromosome and show how terribly stable it is. Whereas I think, what does it matter?'

Graves is confident her gloomy prophecy won't spell the end of mankind. She's certain that the human male would simply evolve a fresh genetic trigger for his gonads. Other mammals have already done so. A spiny rat from Japan (*Tokudaia osimensis*) and a Transcaucasian mole vole (*Ellobius lutescens*) are just two species of mammal known to have completely lost their Y chromosomes, yet hung on to their testicles. Both males and females have a solitary X chromosome and sexual development is triggered by an entirely different, and as yet unidentified, master sex-determining gene.

Fresh chromosomal oddities are turning up all the time amongst obscure little brown rodents. In South America there are nine species of vole from the genus *Akodon* in which a quarter of females are XY, not XX. Their Y chromosome is complete with SRY, yet they still develop ovaries and produce viable eggs, suggesting they must have an entirely new master switch gene that can suppress the bossy SRY.

These peculiar rodents with their perverse sex chromosomes appear to be an evolutionary botch job. And Graves agrees: they basically are.

'If you or I were designing a creature, we would never come up with something that stupid,' she explained. 'But that's what evolution came up with. And the only way you can explain it is that it evolved from another system and it had advantages. Even if we don't know what those advantages are.'

Now in her eighties, Graves has spent her career investigating the evolutionary genetics of sex in an astonishing array of animals, and still fizzes with enthusiasm for her subject. She's now 'sliding back the evolutionary scale' and studying ancient creatures like the lancelet, *Amphioxus* – a primitive fish with no backbone – and even nematode worms. And, to her amazement, she keeps finding the same old genes cropping up in similar sex-determining pathways, albeit with different triggers. 'These genes have been hanging around for a long time. They've been doing something about sex, not necessarily the same thing, but they're there. I find that quite hair-raising,' she professed, her eyes glinting.

Sex is a master at reinventing itself. It has to be. It is essential in order for sexually reproducing species to persist, after all. This anarchy of common genes may have once been more logical and linear, hundreds of millions of years ago at the start of sex. But eons of evolutionary time have left their mark, creating an extraordinary array of apparently nonsensical, yet somehow functional, botched systems in this ever-evolving sex-defining chaos.

'Nothing makes sense, except in the light of evolution,' Graves offered, wisely, quoting the infamous words of Theodosius Dobzhansky, father of Physiological Ecology. 'You have to get over the idea that this was meant to be. Nothing is meant to be. We're all being buffeted by the forces of evolution all the time.'

The confusion of sex chromosomes seen in mammals is just the tip of the iceberg when it comes to the bewildering diversity of systems that exist in nature. For a start, not all sex determination follows the genetic XY system. Birds, a number of reptiles, and butterflies have much the same sex-determining genes, but on different sex chromosomes – a big Z and a withered W. In this system the reverse pattern is the norm – females are ZW and males are ZZ. In this alternative ZW system the master switch gene may be highly conserved, as SRY is in the majority of mammals, or vary between closely related groups.

In some reptiles, fish and amphibians, sexual differentiation might not be triggered by a master sex-determining gene at all but instead stimulated by an external factor. Turtles, for instance, haul themselves out of the sea to bury their eggs in the sand on tropical beaches. Eggs incubating above 87.8 degrees Fahrenheit will activate genes to create ovaries, whereas those below 81.86 will make testes. Temperatures that fluctuate between the two extremes produce a mixture of male and female baby turtles.

Heat is just one of several known external sex-determination stimuli. Exposure to sunlight, parasitic infections, pH levels, salinity, water quality, nutrition, oxygen pressure, population density and social circumstance – how many of the opposite sex are in your neighbourhood – can all influence an animal's sexual fate.

In some animals sex determination can be controlled by any, or indeed many, of the above. Which means sex can get very confusing indeed if, for example, you are a frog.

Nicolas Rodrigues might just have the best job in the world. He spends spring in the Swiss Alps hanging about high-altitude ponds surrounded by snow-capped mountains and verdant pastures scattered with wild flowers and the occasional goat herd – an idyll straight out of *Heidi*. This evolutionary biologist's job is to catch frogs: tiny baby common frogs, *Rana temporaria*, that have just metamorphosed and are graduating from their pond nursery to an adult life on land. Sometimes he has to wait for days and just drink in the view until, all of a sudden, an army of the little hoppers emerges en masse and it's time to get busy with his net.

If he ever needs an assistant, I'm his woman. I spent some of the happiest days of my childhood catching common frogs in a pond a few fields from my parents' house. Like Rodrigues, I was fascinated by the cute little metamorphs bouncing out of the pond. To me they represented the pioneering evolutionary explorers that forged the great leap from water to land some 400 million years ago. Inside the bodies of these emerging froglets an almighty upheaval of tissue and organ reconfiguration means they must switch to obtaining their oxygen by breathing air through budding lungs, rather than filtering it out of water using gills. Many would emerge still clutching a souvenir of their aquatic youth in the form of the unabsorbed tip of tadpole tail, suggesting to me they might also be exiting the pond with their vital air sacs a touch undercooked.

It turns out these adolescent amphibians were more liminal than I could have imagined. Around half of the froglets I caught would also have been in the throes of another major organ change – their ovaries would be switching to testes as they transitioned from life as an aquatic female tadpole to that of a terrestrial male frog.

Sexual differentiation isn't exactly a watertight process if you're a common frog. In fact, according to Rodrigues, it's more than a bit 'leaky'. He's part of a team that has discovered the master switch for these frogs to develop testes rather than ovaries is sometimes genetic, sometimes environmental, sometimes a bit of both. It all depends on where the frogs are from.

The common frog is widespread throughout Europe, from Spain to Norway. These familiar little brown amphibians are all the same species but according to Rodrigues they fall into three different 'sex races' depending on their mode of sex determination.

Common frogs from the northernmost parts of their range have the familiar XY genetic sex determination and develop as one would expect – XY individuals grow testes and XX develop ovaries.

The frogs I caught as a kid were in the southern range and their sex is a little more fluid. All tadpoles are XX and develop as females. But as they emerge from the pond, around half of these genetic females reverse their sexual development. Their ovaries transform into testes and they become XX males.

Switching sex might seem like a big deal, but frogs do it without batting an eyelid (or rather, I should say 'eyelids', since they are in possession of three for each eye). The underlying mechanism isn't fully understood but is thought to be temperature-related. In the laboratory, frogs have been encouraged to change sex from male to female by exposure to chemicals that mimic oestrogen. These are found in herbicides like Atrazine, popular with lawn-growers in the USA whose liberal use of them forces male frogs to switch sex and become female.

The frogs in the middle range are intermediates in every way. Some males have their sex governed by temperature and start out with ovaries; others are triggered by sex-determining genes. As a result, some frogs are regular XY males and XX females but Rodrigues has also documented XY females and XX males. Externally these frogs appear to be either male or female, but their gonads tell a different story. Some have a blend of ovarian and testicular tissue, which makes marshalling their sex into one of two neat buckets all but impossible.

'There is a continuum between male and female at the gonadal level and at the genetic level, but if you go to a random pond and catch a random frog it will still look like either a male or female,' Rodrigues told me.

It would be easy to dismiss this sexual mishmash as the glitches of an imperfect, less-evolved system of sex determination. Many scientists have. But that's a primitive, mammal-centric viewpoint. This extraordinary plasticity is now understood to persist in a range of reptiles, fish and amphibians. It's persevered for hundreds of millions of years across diverse species, which suggests there must be some evolutionary benefit.

A recent study on central bearded dragons (*Pogona vitticeps*), an Australian desert reptile with an impressive spiky neck, gave clues to this benefit. Researchers discovered that the combination of environmentally triggered sex reversal and genetic sex determination has the power to create two distinct types of female.

Most bearded dragons have genetic sex determination – females develop from ZW sex chromosomes, and males from ZZ. But this genetic sex-determination system can be overridden by excessive heat. If during development a clutch of ZZ male eggs gets baked by too much Australian sun, the high temperature overrides their chromo-somal sex and ZZ males switch sex to female.

These sex-changed ZZ females have a unique constellation of male-like and female-like physical and personality traits. They lay twice as many eggs, yet their behaviour is more in line with male dragons – they're bolder, more active and their temperature is higher. This novel variation allows genetic or sex-reversed female dragons to respond differently to a more diverse range of environ-mental pressures, giving them an evolutionary advantage.

The researchers behind the study noted that although the dragons' gonads may be female, their behaviour and morphology are more masculine, leading them to propose that these sex-reversed super-charged dragons should perhaps be considered a separate third sex – one which could offer the species distinct fitness advan-tages. Rather than being seen as 'an aberration', this hotchpotch of sex-determination systems and the resulting sex-reversed females could, in fact, be a powerful driver of evolutionary change.

These sex-reversed dragons, with their mixture of female gonads and male behaviours, also throw a spanner in the Organizational Concept. Their 'male-like' brain appears to be driven by their inher-ent genetic make-up, rather than the cascade of hormonal changes initiated by sex determination. They are not alone. In the last few decades, research into other sexually ambiguous animals has chal-lenged this universal paradigm of sexual differentiation and begun to reveal the extraordinary complexity of sex and its expression in gonads, bodies and brains across the animal kingdom.

In 2008, a retired high school teacher named Robert Motz was staring out the window into his backyard when he spotted a rather curious bird. One side of the bird's body was covered in striking scarlet feathers and topped off with a dramatic red crest, while the other side was a dowdy buff brown. It looked as if two half birds had been glued together down the middle, and, in a way, they had.

The bird was a gynandromorph – an exceptional intersex that's split straight down the centre line. The showy red side was a male cardinal bird complete with solitary internal testicle, whereas the brown side had an ovary instead. This bilateral condition is rare but has been documented in a number of birds, butterflies, insects and crustaceans – animals all with the ZW sex-determination system. They're particularly spectacular in sexually dimorphic species like the cardinal, and arise when fertilized twin embryos fuse very early during development – between the 2-cell and 64-cell stage – to form a chimera with ZW sex chromosomes (female) on one side and ZZ (male) on the other.

These 'half-siders' offer a unique opportunity to test the authority of gonadal sex hormones in shaping brains and behaviour. Gynandromorphs may be made up of two sexes, but they share just one bloodstream, which means they're bathed in the same hormonal milieu. Is the solitary testis and its brawny androgens the supreme driver of sexual fate for the chimera's entire brain, as the Organizational Concept predicts, or could the 'passive' feminine side somehow prevail?

One of the first 'half-siders' to fall into scientific hands was discovered in a physician's poultry yard in Canada in the 1920s. Dr Schaef noticed one of his chickens looked like a hen from one side and a rooster from the other. This funky chicken's behaviour was equally confused: the cock tried to mate with the hens but also laid eggs.

Unfortunately, before the bird's brain and behaviour could be fully scrutinized the good doctor took the unconventional move of killing this valuable anomaly and roasting it for supper. Schaef donated the bones and eviscerated gonads to an anatomist friend who noted in great detail how the skeleton on one side of the bird was bigger and more cock-like, but how the chicken's ovaries, although functional, contained some testicular tissue. She imagined the mixture arose from a conflict of male and female hormones produced by the dual sex organs, but could presume no more on account of most of her study subject having been eaten by Dr Schaef.

Almost a century later Arthur Arnold, a research professor at the University of California, Los Angeles, got his hands on a zebra finch gynandromorph. He chose not to eat it, but instead eagerly examined the bird's brain. Zebra finch are songbirds, but only the males sing, so their neural circuitry is more developed than females'. This zebra finch had been observed singing, so Arnold assumed it would have a uniform 'male' brain. When he dissected the bird, however, he discovered the female side of the brain a little more masculinized than normal, but crucially the bird's song circuit had only developed on the gynandromorph's male side.

'It blew me away,' Arnold told *Scientific American* at the time. The gynandromorph's semi-female brain cast doubt on the omnipotence of gonadal steroids to differentiate sexual dimorphism in birds. In other words, this bilateral intersex bird kicked the Organizational Concept in the nuts. Here was evidence that androgens were not the exclusive force sculpting the sexuality of a bird's body, brain and behaviour. Instead the sex chromosomes, exerting their identity inside neural cells, must be playing a crucial role.

Gynandromorphs can also develop as sexual mosaics, with ZZ and ZW cells intermingled throughout the body rather than being organized as a neat bilateral hermaphrodite. A later study of three such gynandromorph chickens found that cells throughout the birds' bodies followed their own sets of genetic instructions, and were not necessarily dominated by the sex hormones to which they were exposed. So, with birds at least, the genetic sexual identity of individual cells plays a significant role in driving sexual dimorphisms in the body and brain.

'Sex is not a unitary phenomenon,' David Crews explained to me over the phone. The recently retired professor of zoology and psychology at the University of Texas should know. Crews spent forty years unpicking the mechanisms behind sexual determination and differentiation in an eclectic cast of wild animals. He's decoded the exact genes involved in gonad development in turtles, encouraged whiptail lizards to switch sex, and monitored how incubation temperature affects not just the sex but also the sex appeal of leopard geckos.

According to Crews there are five types of sex: chromosomal, gonadal, hormonal, morphological and behavioural. They don't necessarily all agree with one another or even remain fixed for life. They are cumulative and emergent in nature, and can be influenced by genes or hormones, as well as the environment or even an animal's life experience. This plasticity allows for the huge variety in sex and sexual expression that we see both within and between species.

'Variation is the fabric of evolution. If you don't have variation you can't have an evolving system. So it's important that we have variation in sexual characteristics.'

Crews is a self-confessed free thinker whose fresh perspective comes from studying wild reptiles, birds and fish as opposed to laboratory-bred mice – the standard animal archetype for studies of sexual development. These unconventional model organisms, he tells me, are 'real' rather than simply 'realistic' – their natural instincts have not been blunted by decades of inbreeding. Their sexual development is triggered by an array of factors – genetics, temperature or environment – giving him the opportunity to look beyond the standard lab mouse model and travel back in evolutionary time to study the systems that existed before mammalian sexual development, yet formed its basis.

Crews blames the Organizational Concept for promoting a rigid deterministic view of sex, which focuses on the differences between the sexes, reinforcing the binary concept and ignoring the glorious diversity of sexual characteristics found in nature.

'It's offensive,' he spat down the phone from his home near Austin during one of our many long and fascinating chats. The standard paradigm has, in his opinion, had its day. It is mammal-centric, overly simplistic and underplays the role of oestrogen as an organizing and activating sex hormone. 'Females are just as differentiated [active] as males. I've tried to make this point several times. My conclusion has been that the female is the ancestral sex. I think there's a lot of evidence for that.'

Crews has focused his career on studying the diversity itself and how it is actually controlled by the same mechanisms. Studying what is conserved in all this botched chaos is the key to discovering what is fundamental. This approach has enabled Crews to develop an alternative evolutionary perspective for thinking about sexual differentiation. One which is grounded in the very origin of sex.

'There is little doubt that the first creatures reproduced by cloning,' he told me. 'The earliest reproductive organism had to be able to lay eggs and that's a female.'

Crews' research estimates that 600–800 million years ago the only creatures in existence were these cloning egg-layers. Males did not arrive on the evolutionary scene until the dawn of sex, when gametes diverged in size, which Crews reckons was around 250–350 million years later. With this divergence came the need for complementary behaviours to facilitate the union of these different-sized gametes; individuals must locate one another, become sexually attracted and reproduce. So sexual dimorphisms evolved that were activated by androgens.

'Maleness evolved as an adaptation to femaleness,' Crews continued. 'When males came along what they did was to facilitate reproduction in the female. To stimulate and coordinate the neuro-endocrinological processes that underlie the shedding of gametes. Males are behavioural facilitators.'

If males are the derived sex, that evolved out of the original female, it is logical to assume they must contain evolutionary traces of egg-makers. And, it turns out, they do. Crews discovered active relics of ancient femininity in the very seat of masculinity: the testes.

'We published the first photomicrographs showing that the testis is loaded with oestrogen receptors,' he told me. Oestrogen, the primary 'female' sex steroid hormone, turns out to play a fundamental role in the development of male testes and sperm.

Crews collaborated with Joe Thornton, a professor of genetics at the University of Chicago, to do some molecular time travel and resurrect the ancestral receptor for oestrogen from an ancient mollusc. Thornton's work on this, and other primitive animals like lampreys, has subsequently shown that the oestrogen receptor is the oldest transcription factor (a protein whose job is to turn genes on or off) in vertebrates – far older than previously thought, with an origin between 600 million and 1.2 billion years ago. The genes for androgen receptors did not evolve for a further 350 million years.

'Oestrogen has to be the original steroid hormone because the ancestral animals only produced eggs, and eggs produce oestrogen,' Crews explained. 'The oestrogen receptor is important in virtually every tissue of the body. I can't think of any tissue in the body that doesn't have an oestrogen receptor.'

The Organizational Concept may have focused on the omnipotence of testosterone, but oestrogen is proving to be equally powerful. It has even been demonstrated to have the same organizing effects as testosterone in early development, with the ability, as we have seen, to cause sex reversal in frogs. Crews has also reversed the sex of developing female lizards using oestrogen blockers. Oestrogen clearly plays a fundamental role organizing both female and male sexual development but also activating sexual behaviours later in life. Not only is the 'female' sex hormone required to make testis and sperm, it is also understood to stimulate male copulatory behaviour in some species.

'The "female" sex steroid has a critical role even in males, because males were originally females,' Crews expounded.

So, in the gospel according to Crews: Eve wasn't created out of Adam's rib, it was the other way round. In the beginning there was female, and she gave rise to male. From this alternative evolutionary perspective, the ultimate answer to *what is a female* is: she's the ancestral sex. Relics of this primal egg-layer exist within all of us. Which puts a fresh spin on males getting in touch with their feminine side.

CHAPTER TWO

The mysteries of mate choice: robo-bird to the rescue

Few animal courtships are as strange or, quite frankly, as silly as that of the greater sage grouse (*Centrocercus urophasianus*). This North American bird, about the size of a large chicken, ekes out a frugal existence on the Great Western Plains eating – you guessed it – sagebrush. In early spring, sage grouse bachelors – resplendent in their spiky fantail finery – gather in large numbers at designated patches of sagebrush prairie to compete for a mate. Known in zoological circles as leks, these events are essentially sage grouse discos; the battle for sex is played out using the medium of dance, with the males strutting about, providing their own unlikely soundtrack by, basically, beatboxing.

Male greater sage grouse have a massively distended oesophagus which they can inflate by gulping down mouthfuls of air to create a large wobbly white-feathered throat balloon, which when fully swollen briefly exposes two bulbous patches of olive-green skin that pop forth from their feathers like a pair of nippleless shop-dummy breasts. It's a pretty eye-catching look and, by controlling the expulsion using an impressive web of chest muscles, the cocks are able to slap their olive air sacs together to generate an even more arresting noise: a loud, high-pitched *doink* that sounds as if it were made by twanging rubber bands over water.

The overall effect is pure Monty Python and begs the question,

Evolution, what were you thinking? What perverse force could have shaped such preposterousness? The answer: female choice.

Female animals have rather a lot to answer for. Why does the male proboscis monkey have such a long and pendulous nose? Well, the ladies apparently like it that way. Ditto the stalk-eyed fly's unwieldy horizontal eye-stalks (which stretch wider than its body is long) and, of course, the greater sage grouse's body-popping strut. Female choice is the most whimsical of evolutionary powers, with a hand in some of nature's most extravagant creations. Trying to fathom exactly what females are choosing and how has become one of the most dynamic areas of evolutionary biology in recent years, and has generated insights using methods that are sometimes as surreal as the sage grouse themselves.

One of the leaders in this field is Gail Patricelli, a sparkly-eyed young professor of evolution and ecology at University of California, Davis, who has been studying the greater sage grouse strut for the best part of ten years. Before I went to meet Gail at her lab, she generously hooked me up with a coveted ticket to the sage grouse show by connecting me with one of her graduate students, Eric Tymstra, who's been helping her conduct a long-term study of these strange birds in California's Eastern Sierra. Eric and I corresponded via email to arrange our rendezvous at nearby Mammoth Yosemite Airport.

'How will I recognize you?' I asked, a little anxiously. I need not have worried. 'I'm wearing teal and I have a Mohawk,' came Eric's blunt reply.

As promised, Eric was as exuberant as his study subjects: easy to find and, with a fiendish sense of humour, easy to like. On the drive to the field site he laid out my options for birding. I could either get up at one a.m. and join him trapping and tagging birds, or have a 'lie-in' and wake at four a.m. to simply watch them dance on the lek.

'What does tagging involve?' I asked. First, Eric explained, they must find the bird, which they do by scanning the sagebrush with a flashlight and looking for a pair of reflective retinas gleaming in the dark. Sage grouse are 'pretty dumb' so the light will temporarily stun the bird, allowing Eric and his colleague a small window to march forward and grab it with a net. 'Most people blast white noise to disguise the sound of their footsteps as they approach the bird,' Eric told me, 'but I use AC/DC.' How could I resist?

And so I came to spend my first night in the mountains trudging for several miles through thick snow in sub-zero conditions, trying to keep up with Eric and his equally athletic colleague as we scanned the brush for bright beady eyes. I was a sluggish companion, shuffling along in a dozen borrowed layers like some sort of unwieldy thermo-padded human pass-the-parcel, unused to the altitude (roughly 9,000 feet) and the ill-fitting snowshoes, which kept tripping me up and forcing me to face-plant into the unseasonal drifts of thigh-deep snow. Given the lead weight of my presence it is not surprising that we failed to catch a bird that night, but Eric was such a sweetheart he didn't make me feel bad about it.

At around five a.m., while it was still pitch dark, we cut our losses and squished into our tiny two-man bird hide to set up a plethora of lenses: binoculars, monoculars and multiple cameras to observe and record the action on the lek. 'We're basically *pornithologists*,' Eric joked. 'It's our job to record bird sex.'

The avian orchestra began limbering up pre-dawn, teasing its audience with the eerie sound of echoing *doinks* as the sky turned from black to blue. By the time the rising sun had painted the surrounding snow-clad mountains pink, I was able to make out a bunch of black shuffling blobs in the distance. The show had begun and it would not disappoint.

The scene that played out in front of the hide was every bit as bizarre as I had hoped for. About thirty male sage grouse were body-popping on a patch of ground roughly the size of two netball courts. The surrounding sunrise-tinted mountains provided a natural amphitheatre for their performance, enabling the sound to reverberate up to three kilometres, broadcasting their presence to the opposite sex. To begin with, there were no females on the lek, but this did not deter the males, who danced away regardless, apparently lost in their own world of solo performance. It looked like an awful lot of effort for no reward, though clearly the males were paying attention to one another. Every now and again a fight would break out: a flurry of surprisingly violent wing-beatings would see one male silenced while the winner did a body-popping glory parade in front of him.

I'm not alone in enjoying the bird's absurdity. 'One of the things I love about studying these birds is they take themselves so ser- iously,' Gail confessed to me in her lab at UC Davis. 'It's just totally ridiculous what they are doing – it's obscene – and yet it's deadly serious for them. This is the crux of evolution. This is when the genes get passed on to the next generation. All the surviving and avoiding predators means nothing in evolutionary terms unless they can mate and so it really is ground zero evolution.'

When the females did show up it only added to the comedy. The presence of the smaller and decidedly more dowdy hens triggered the males to ramp up their giddy display several gears. Yet the females themselves could not have looked less interested. They hud- dled in loose groups, occasionally pecking at the ground in an indiscriminate fashion, apparently oblivious to the frantic postur- ing and popping going on all around them. The males compounded this desultory scene by being tragically off beam with their perform- ance; repeatedly turning their backs on the females and directing their sexy sac-slapping away from the bored-looking hens.

'One of the things that got me interested in sage grouse is that it looks like all your Victorian stereotypes come to life, right?' Gail said. 'The flashy aggressive males are fighting with each other. They're putting on a show and the females are passively playing coy.'

Descriptions of sage grouse leks have traditionally embodied androcentric typecasts. The birds made their dramatic debut on the ornithological scene with a male displaying his fully inflated sacs on the cover of *Nature* magazine in 1932. The author of the paper, R. Bruce Horsfall, took great delight in describing the males' 'queer antics' but assumed their 'rubbery plops' must be directed at one another, and not the females. This view was upheld for most of the twentieth century, with scientific papers entering into lengthy discussions about the dominance hierarchy of the 'master cocks', with the hens' 'inconspicuous and passive' behaviour demanding little attention.

'That really was a reflection of this view that it's got to be all about the guys,' Gail explained. 'All this communication was just about males threatening other males and it's not driven by female choice.'

Female choice may be one of the hottest subjects in evolutionary biology today, but it hasn't always been that way. Darwin first proposed the evolutionary power of female fancy as part of his theory of sexual selection, which he outlined in detail in *The Descent of Man, and Selection in Relation to Sex*. Darwin's second great evolutionary principle was intended to plug some troublesome holes in his theory of natural selection, namely those created by the outlandish adornments and sexual displays of certain male animals.

In a now famous letter to Asa Gray in 1860 – the year after *On the Origin of Species* was published – Darwin confessed, 'It is curious that I remember well the time when the thought of the eye made me cold all over, but I have got over this stage of the complaint, & now small trifling particulars of structure often make me very uncomfortable. The sight of a feather in a peacock's tail, whenever I gaze at it, makes me sick!'

Darwin's queasiness concerned the unbridled frivolity of the peacock's tail, which didn't appear to benefit the bird's overall quest of survival. In fact it seemed more likely to have a negative impact by impeding the peacock's ability to hide or fly away from danger. So how and why could such excess have been shaped by the utilitarian force of natural selection?

Darwin's revolutionary proposition was that such 'secondary sexual characteristics' could be explained by two forces: male–male competition for females, which led to armaments like the outsized horns of a rhinoceros beetle; and female mate choice, which shaped ornaments like the peacock's tail.

'It is shown by various facts ... the female, though comparatively passive, generally exerts some choice and accepts one male in preference to others.' Later, Darwin goes on to outline the effect of such whimsy: 'preference by the female of the more attractive males would almost certainly lead to their modification; and such modification might in the course of time be augmented to almost any extent, compatible with the existence of the species'.

The Victorian patriarchy had little trouble swallowing the idea of males duking it out over the right to mate with a female being a potent evolutionary force, although most argued it was simply a subclass of natural selection. Darwin's controversial claim was that females were not only sexually autonomous but had the wherewithal to make decisions that shaped male evolution. This put the fairer sex in a very powerful role – a role that made most (male) biologists deeply uncomfortable. Men controlled women in Victorian England – not the other way around.

The astonishing originality of Darwin's new theory didn't help its passage into cultural and scientific acceptance either. Whilst the theory of evolution by natural selection had been anticipated by many eighteenth- and nineteenth-century thinkers, and was co-discovered by Alfred Russel Wallace, the notion of sexual selection as an evolutionary force was without scientific precedent. To make matters worse, Darwin did not propose any kind of adaptive explanation for female mating preferences. Instead he attributed female attraction to a 'taste for the beautiful'. Although he enters into lengthy analysis of which animals had sufficient mental capacity to make such decisions (insects yes, worms no), Darwin gave the impression that animals required a human-like sense of aesthetics for sexual selection to work.

This gave the Victorian establishment a whip to beat Darwin's new theory with. According to the thinking of the time, only the upper classes could appreciate art or music, so it seemed utterly absurd that a female, let alone a lowly peahen, would be endowed with an aesthetic faculty. Beauty was God-given, so the idea that female sexual preference was the primary agent of its evolution was tantamount to heresy.

Darwin's audacious new theory was openly mocked and dismissed. His most influential detractor was Alfred Russel Wallace, who saw no need for a bogus new evolutionary theory to explain male courtship ornaments and display, which he believed were simply a result of the male's 'surplus of strength, vitality, and growth-power'. In his unambiguously titled evolutionary tome, *Darwinism* (1889), published a few years after the great man's death in 1881, Wallace took the bold move of censoring Darwin's legacy. 'In rejecting . . . female choice I insist on the greater efficacy of natural selection. This is pre-eminently the Darwinian doctrine, and I therefore claim for my book the position of being the advocate of pure Darwinism.'

Wallace may have lost out on equal credit for the theory of evolution by natural selection, but he won the war when it came to modelling Darwinist thinking in the twentieth century. His trashing of female choice meant that Darwin's second great theory of sexual selection became 'the mad aunt in the evolutionary attic of Darwinian theory', which, apart from a couple of exceptions, wasn't allowed out to play for one hundred years. When the major evolutionary biologists of the twentieth century discussed extravagant male traits, they were thought to frighten away predators or help females to find a mate of the right species.

Things have changed. The sexual revolution of the 1970s and the impact of feminism on evolutionary biology have helped stir Darwin's daring idea from its century-long slumber. The idea that females – from birds to fishes to frogs to moths – are able to make sensory evaluations and exercise mate preference has been scientifically proven and accepted. Numerous studies have demonstrated how females in diverse species prefer brighter colours, louder calls, stronger odours and faster dances. In the last thirty years female choice has become one of 'the most dynamic areas' of evolutionary research, with lekking species such as the greater sage grouse providing the paradigm case for competitive males and choosy females.

Leks are the most extreme seduction marketplace – a winner-takes-all situation where a lucky few bachelors will dominate the mating scene: 70–80 per cent of copulations on leks – from insects to mammals – are attributed to only 10–20 per cent of the males.

'During the peak of mating the sage grouse lek is a madhouse,' Gail Patricelli explained during my visit to her lab. Most of the mating happens in just three days, with the females often in one big scrum, fighting with each other for their position around the top cock. Patricelli regaled me with the legend of Dick: a dominant male sage grouse in Wyoming who mated 137 times over the 2014 season, with twenty-three of those copulations taking place in twenty-three consecutive minutes.*

What makes a lek so special for students of sexual selection is that the females will go on to raise their offspring alone. So they're not making their mate choice based on the resource richness of a male's territory or his potential parenting skills – they are simply after his genes. Given that the winning male will pass the genes that underwrote that display on to most of the females' offspring, the power of female mate choice to shape outlandish male ornaments and courtship displays is unbridled.

'Dick was the one that passed on almost all of the genes in that season. And so there's very strong selection to be that guy,' Gail explained. 'Lekking animals are doing some of the craziest things that you'll see out there in the natural world because the selection is just so over the top. They are the reason why you see birds of paradise and peacocks and the sage grouse strut.'

But the million-dollar question is, what made Dick so alluring?

---

* Gail and her team name the birds after the patterns they trace via the white tips to their tail feathers, which are as unique as a fingerprint. Dick's feathers resembled the shape of a penis, so they christened him Dick without knowing the extent to which the dominant cock would live up to his name.

'Dick was actually quite extraordinary,' Gail told me. 'He was strutting so hard all the time. It just seemed like he had endless energy.'

The sage grouse's beatboxing strut looks as strenuous as it is bizarre. How much energy it costs the cocks is hard to say, but a recent study on great snipe, another lekking species, that courts females by drumming its wings, found they lose almost 7 per cent of their body mass after every daily courtship session. So there are likely some serious energetic costs for the male sage grouse, especially since their sagebrush is so low in energy. On top of that, Eric told me that the shrub's leaves are also highly toxic, so the males are essentially dancing their guts out while coping with the avian equivalent of a massive hangover.

Gail thinks that by demanding the cocks perform this gruelling dance the hens ensure that they bag themselves a high pedigree male, with grade-A genes. 'There's not one single system that's involved with making somebody a world-class athlete versus someone who isn't, right? It's going to be your aerobic capacity, your metabolic efficiency, your immune system, your foraging ability, how well you digest food and turn it into active energy, and so on. So the sage grouse are working their butts off doing this fundamentally difficult thing that you can only do if you are in good condition.'

It would be easy to assume this is the end of the story – the flashiest male with the greatest stamina wins the female. And many scientists have. But Patricelli was intrigued by the apparently coy nature of the females. Were they really as passive as they seemed?

To find out, Patricelli took inspiration from colleagues at the Cornell Lab of Ornithology, Marc Dantzker and Jack Bradbury, who'd made a remarkable discovery: the acoustics of the males' beatboxing had a weird four-pronged pattern of sound radiation that means their calls are actually quietest directly in front of the displaying bird and loudest to the sides and behind. So although the strutting male appears to be turning his back on a female, he is actually blasting his rubbery plops directly at her. This made Gail think – if the male's display is not what it seems, then maybe the hens are not as boring as they appear either.

By turning her attention away from the eye-catching display of the males and focusing on the modest hens Gail discovered something even more revelatory. Top cocks like Dick are not simply the loudest dancers on the lek, they also need to respond to subtle cues given by the females. Meaning they need to *listen* as well as they can dance.

'What I have been looking at is the way there is this interplay in communication going on between the males and females. The female is playing a much more active role – either eliciting the male display she wants or shaping the male displays – and he has to respond in order to be attractive and not just be flashy. And that's where the robots come in.'

To really get inside the hens' heads and figure out what choices they are making, Gail created perhaps the only bird even more surreal than a beatboxing cock: a robot female sage grouse. Gail fashioned her 'fembot', as she affectionately likes to call her, out of a taxidermy skin of a female sage grouse, a remote-controlled tank kit and some robot parts she bought online, all held together with a pair of Spanx. Thanks to Gail's handicraft skills the resulting homespun robo-bird was actually incredibly realistic, apart from the wheels.* These don't seem to bother the males, however, who are an indiscriminate bunch. Gail has seen them attempting to mate with dried cowpats when there are no females around – the bar is set pretty low evidently. Nevertheless she was nervous the males might not like her fembot the first time she drove her into the lek. 'It had a very odd first date quality to it,' she told me.

I got to take Gail's fembot for a spin around her lab. The bird

---

* Gail credits her year and a half off as a 'ski bum in Colorado' as inspiration for creating robo-birds. To earn her keep she had a sideline in organizing conferences, and neuromorphic engineering – all about AI and robotics – was her favourite. So when she had this idea she was able to reach out to her conference buddies, one of whom just happened to design control systems for NASA and was happy to take on her robo-bird as a side project. 'It was a very sophisticated tool in that first iteration,' she told me. It was, however, still held together by Spanx. 'That's why I always tell students to take time off before they start grad school because you never know what you're going to see or learn.'

can be directed to flirt or be coy on command by approaching, turning her head and directing her gaze or putting her head down and foraging – just like a real hen. Controlling the fembot in the lab on a smooth open floor is significantly easier than on a lumpy, crowded lek. This takes two people: Patricelli is tucked away in a bird hide with the controls while one of her PhD students is perched on a hill with binoculars and a radio, helping her navigate the lek. 'It's like this really stressful game of Frogger,' Gail admitted. 'The actual birds are moving all over the place and you're trying to move amongst them without having anyone knock the robot over or crash into anybody.'

Gail's fembot has to be the ultimate tease. She must arouse the interest of a male, but duck out before things get too hot and heavy. 'We have had males try to mate with the robot and that does not go well.' Nevertheless, accidents do happen. Once a fembot lost her head mid-flirt, which was awkward, but fortunately the dumb cock wasn't bothered. 'The males don't know how to react as they don't recognize it as something, so I just drove her back headless.'

Controlling the female side of this complex courtship conversation has enabled Gail to observe how changes in the hen's behaviour affect the male's performance. 'We see that the males are adjusting the rate of their display according to the proximity of the females. They're actually responding and using their energy where it would matter most. Unsuccessful guys blast away at top level all the time and then when it comes to crunch they don't have much left to put on a great show. And that's probably a combination of social skills and their own underlying health.'

Gail first discovered the importance of these 'social skills' in one of the most sophisticated of all lekking males: the satin bowerbird of eastern Australia, *Ptilonorhynchus violaceus*, which she studied for her PhD. Male bowerbirds are the animal kingdom's closest thing to Salvador Dalí, creating fanciful surrealist bowers out of sticks, which they then go to great lengths to decorate with brightly coloured objects that please the female's eye. The style and colour of ornamentation varies according to the species, with some bowers being more elaborate than others.

The great bowerbird's bower, for example, is like a house of illusions: objects are arranged according to their size to create a false perspective that makes the bower seem smaller and the male seem bigger than he really is.*

Female satin bowerbirds are less fussed about the size of their cock and more concerned about his ability to acquire blue trinkets. Males hunt high and low to scatter the floor of their bower with anything blue they can get their beaks on – from flowers to feathers to plastic bottle tops and clothes pegs. As if this weren't surreal enough, they have even been known to paint the walls by holding a bit of bark coated in chewed up berries in their beak. Dalí would be proud.

Once the males have got the hen's attention with their blue bling, the cocks show off their physical prowess by puffing up their shiny azure feathers and doing an elaborate buzzing dance. This isn't that dissimilar to the display fighting males use to intimidate each other, so it's not surprising that females are often more than a little bit jumpy during early courtship. Gail noticed that more assured females did a sort of crouching motion and she wondered if the males were paying attention and responding to this. So she built a robo-bowerbird that could mimic this move and discovered that successful males were indeed the most tactical and considerate: only ramping up the intensity of their dance when the female crouched down and signalled she was ready for it.

Previous scientists had noted male animals paying attention to female signals during courtship. But Gail was the first to show that listening and responding to her cues was linked with the male's mating success. Not only that, but Gail has calculated that, at least in the case of the bowerbird, these social skills are

---

* Great bowerbirds may be master illusionists but parrots, thrushes, pigeons and even chickens have all been shown to be sensitive to various illusions, and males of many species display themselves to females at a characteristic angle and distance; suggesting that the use of illusions might be widespread in birds.

equally as important as the intensity of his display for the male to be successful.

Female sage grouse may also be paying attention to each other. Young female guppies are known to emulate older (and perhaps wiser) female fishes' mate choice decisions. Gail believes that such 'eavesdropping' may also contribute to the insane allure of dominant male sage grouse like Dick. 'If eighty females are around a male then it's hard to imagine they all made an independent decision.' Back in 2014 Gail tried to get real hens to follow the interest shown by her fembot in a non-dominant male. But she was thwarted by the exceptional charisma of the dominant male that year: 'In the end, we could not overcome the magnetism of Dick.'

Patricelli has not given up. Although she admits that building a male sage grouse is beyond her robo-skills, she'd like to build more fembots to help tease apart the dynamic tactics of males and females on the lek. Courtship has traditionally been viewed as a black box in which males and females assort themselves according to the quality of a male's trait and the strength of a female's preference, with the process itself seen as obscure or irrelevant. Gail sees a lek more like a big open bazaar full of traders and buyers in a constant state of shopping and negotiating. Just as Darwin was influenced by the economist Thomas Malthus* when formulating his theory of natural selection, Gail has turned to economic models of negotiation to provide a conceptual framework that emphasizes courtship as a process in which the male and female bargain to reach a deal, influenced by the haggling of other players in the market.

---

* Thomas Malthus was a British social economist best known for his treatise on population growth, which stated that people will always threaten to outrun food supplies unless reproduction is checked. This idea was hugely influential to Darwin as he was formulating his ideas about what force could possibly drive evolution. Before reading Malthus, Darwin had thought that living things reproduced just enough to keep populations stable. But, after reading the economist's work, he came to realize that, as in human society, populations bred beyond their means, leaving survivors and losers in the effort to exist. And thus this competition for survival became the driving force of his theory of evolution by natural selection.

'Sexual selection can drive the evolution of these flashy traits, but also this kind of social intelligence, these courtship tactics that are also an important part of competition for mates. So sexual selection might be a lot more powerful than we initially assumed,' Gail explained.

All this tactical negotiation requires cognitive power. Male satin bowerbirds have relatively large brains, are long-lived and undergo a strange seven-year adolescence which is spent impersonating the female. Juvenile males share the same green plumage as the females and Gail thinks that learning their complex lothario skills might explain this unusually long, cross-dressing developmental period, which is spent not only practising bower-building but being actively courted by adult males. 'Young males learn courtship from the role of the female and they'll often do these crouching displays. They don't "mate", but they'll basically do this entire courtship from the female perspective and then if the male looks like he's getting close to mating they just fly out the front.'

A study in 2009 that tested the male satin bowerbirds' problem-solving skills was the first to show that cognitive performance is associated with mating success and females prefer the most nimble-minded male. Male budgerigars that demonstrate their problem-solving smarts have also been shown to be more attractive to females. So female choice could be responsible for shaping not only a male's body and behaviour but also his brain.

This idea isn't new. Darwin himself proposed that sexual selection could, in fact, be responsible for the exceptional evolution of human cognition – especially the more 'self-expressive' aspects of human behaviour, such as art, morality, language and creativity. The idea that female choice might have polished the human brain into brilliance would have been the ultimate blow to the Victorian scientific patriarchy – hitting them right between the eyes, where it hurt most.

Female choice is indeed a powerful evolutionary force, but also seemingly random. Why do all female satin bowerbirds love Matisse blue, however, and not Turner's yellow?

Darwin's inability to account for the unpredictable, yet nevertheless uniform, nature of female choice provided further ammunition for its detractors. Critics, like Alfred Russel Wallace, found it 'quite incredible . . . that a large majority of females . . . should agree in being pleased by the same particular kind of variation'.

Today many scientists believe the answer to this fashionable fancy lies in the tuning of the female's senses. A male wants to get picked, and this means standing out from the crowd and getting noticed. A sure-fire way to catch a female's attention, it turns out, is by dressing up as her favourite food.

Female freshwater guppies from Trinidad (*Poecilia reticulata*) generally prefer to mate with males bearing larger, more chromatic orange spots. The origin of their preference is thought to be a sensory bias for the colour orange, which has arisen from their penchant for the bright-orange fruits of the cabrehash tree. The ripe fruits drop into their freshwater pools and are consumed voraciously, providing the fish with a vital source of sugars and protein in an otherwise barren environment. So the female guppies have a partiality for the colour orange as it directs them towards a quality food source, which males are subsequently exploiting in order to pique their interest sexually.

It seems the male satin bowerbird could be utilizing the same kind of sensory exploitation to get himself noticed. In experimental situations, the frugivorous females repeatedly preferred blue grapes to other coloured fruit, suggesting their senses may be tuned for alert to this colour. These preferences can run away with themselves over eons of evolutionary time. With thousands of generations of females choosing blue the preference can become seriously distorted. So from a female's fondness for blue fruit you wind up with the male bowerbird's cobalt kleptomania and house of blue booty. It could be that the male sage grouse's olive-green chest is the same hue as the tastiest sagebrush shoots, and his rubbery plops are simply drawing further attention to this existing sensory bias, like a dinner bell sounding for her favourite sage snack.

In the end it is hard to ascertain whether the female sage grouse is picking her lucky mate based on the energy of his dance (and thus the state of his health), the fitness of his genes, his degree of social skills or the simple fact that his chest colour reminds her of dinner and appeals to her 'taste for the beautiful' (or perhaps, the absurd). These concepts are not necessarily mutually exclusive, and the degree to which they apply to any given species is something evolutionary biologists can argue about for days, if not years. Disagreements about whether a female animal gains any actual benefit from her choice, or if it is just aesthetic whimsy driven by a quest for 'hedonic pleasure' are an echoing extension of those originally contested by Wallace and Darwin over one hundred and fifty years ago.

A female sage grouse's mercurial mating decisions are, it seems, no more scrutable than my own. After nearly two decades of intense research the one and only thing that experts in the field agree on is that female choice remains 'essentially mysterious'.

What we do now understand, however, is that mate choice is not a fixed phenomenon. The decision made by a female túngara frog at a mating pond at the start of the night, when surrounded by a raucous cacophony of randy males, has been shown to be quite different to the one she makes at the end. Researchers in Panama found female frogs to be highly selective, showing no interest in synthetic male frog calls made by a portable speaker when the night was young and full of promise. But by the end of the evening, the female frogs were significantly less discriminating. They would happily hop up to the plastic speaker playing fake frog calls and hang about in a rather desperate fashion in the hope of getting their eggs fertilized before the pond party finished.

A female's degree of choosiness might fluctuate according to her age, fertility, environment, life experience or degree of opportunity. Sometimes her choices will involve having sex with more than one male. Sage grouse females may look coy, but they turn out to be surprisingly promiscuous. As we shall discover in the next chapter, a female's choice to mate enthusiastically with multiple partners is enduringly popular throughout the animal kingdom.

CHAPTER THREE

The monogamy myth: female philandering and the great fruit fly fiasco

Hogamus, higamous
Man is polygamous
Higamus, hogamous
Woman monogamous

William James (1842–1910)

I once roared so loud I stole a lion's girlfriend. The roar itself didn't actually leave my mouth, it was a recording of a male lion, played out of a loudspeaker. I was in the Maasai Mara with Dr Ludwig Siefert, a lion specialist, who was demonstrating the use of audio playback in deciphering lion communication. This involved the two of us standing with our heads poking out of the top of his jeep, under the cover of night, pumping the sound of a dominant male's roar into another lion's territory, an audacious way to pursue scientific enquiry and which struck me as the feline equivalent of screaming, 'Come and have a go if you think you're hard enough,' outside a rough pub at chucking-out time.

At first I felt faintly silly projecting our tinny growls into the night. An MP4 and a portable speaker can hardly do justice to the roar of the lion, which at 114 dB is the loudest of any of the big cats. The roar itself is generally less majestic than the one that starts an MGM movie – more of a series of rumbling low grunts – but with the basey resonance to carry for up to five miles. I assumed that our distorted facsimile would arouse little interest. But after a few minutes' silence there came a distant response. Over the next thirty minutes or so we played audio ping-pong with the neighbourhood lion, whose roars became increasingly louder until they made my heart thump, skin prickle and palms sweat.

Out of the gloom emerged not one, but three big cats – two males and a female. All of a sudden our safari vehicle felt rather vulnerable in the face of half a tonne of agitated muscle, teeth and claws. The lions padded around the vehicle, looking for something that looked and smelt like a male lion. When they found nothing, the two males wandered off, but the female lay down in front of our vehicle, pinning us to the spot for over an hour.

Siefert knew these lions. He explained that the males were brothers, and that the female was likely in oestrus and consorting with one of them. Siefert told me the reason she had chosen to abandon her mate and remain by the source of our roar was probably because she was hoping to cop some extra-curricular sex with its owner. Stealing a lion's girlfriend is apparently not that hard. It's not unusual for a lioness to be spotted creeping away from her napping partner in order to engage in saucy trysts with other males. Such wanton duplicity is apparently standard form for the lioness, whose promiscuity is famous amongst big cat researchers – a female lion is known to mate up to one hundred times a day with multiple males during oestrus.

I was shocked and quietly thrilled to discover the licentious nature of the female lion. As the infamous ditty by the philosopher William James attests, everyone knows it is the male of the species that enjoys a profligate sex life, not the female. When I was a zoology student, I was taught that this was the male's biological imperative, written not in stone but in gametes. Anisogamy – the fundamental difference in gamete size, from the Greek, meaning 'unequal' and 'marriage' – is said to define not just the sexes but also their behaviour. Sperm are small and bountiful, whereas eggs are large and limited; so males will be promiscuous and females will be choosy and chaste.

'Excess copulations may not actually cost a female much . . . but they do her no positive good. A male on the other hand can never get enough copulations with as many different females as possible: the word excess has no meaning for a male,' explained my tutor, Richard Dawkins, in *The Selfish Gene*.

This biological law always made my head (and heart) hurt. How could one sex be promiscuous and the other chaste – after all, who were the males having sex *with* if the females were all so demure? It didn't make sense to me. And if a female's sexual behaviour is pre-scribed by her gametes, then how can we explain the unrestrained sexuality of the lioness? Well, it turns out the female lion is by no means the only strumpet in the animal kingdom. The time is long overdue for a radical reappraisal of anisogamy's clichéd sex roles – if only our species is ready to accept them.

THE VICTORIAN MANUFACTURE
OF FEMALE CHASTITY

The female of the species wasn't always pathologically chaste in the eyes of science. At the very birth of zoology, Aristotle noticed domestic hens would routinely get down and dirty with not just one carefully selected cock, but several roosters. It was Darwin who, two millennia later, wiped the sexual slate clean in *The Descent of Man* by forcing the female of the species into an ill-fitting chastity belt.

'In the most distinct classes of the animal kingdom, with mam-mals, birds, reptiles, fishes, insects, and even crustaceans, the differences between the sexes follow almost exactly the same rules; the males are almost always the wooers.'

Darwin's theory of sexual selection, outlined in *The Descent of Man*, continues in the vein of a Mills & Boon romantic novel by stating that male animals have 'stronger passions' and fight amongst themselves to 'sedulously display their charms before the female'. The female, on the other hand, 'with the rarest of exceptions, is less eager than the male. She generally "requires to be courted"; she is coy.' Her job is to passively yield to the winning male's charms, or choose among the 'wooers' and concede to their sexual demands, albeit reluctantly. Darwin noted that the female's demure nature is such that 'she may often be seen endeavouring for a long time to escape from the male'.

Although Darwin did note that in a handful of species the roles were reversed – with females being competitive and males being choosy – he considered these insignificant anomalies. Darwin's explanation for the constancy of his gendered sex roles was that it all boils down to fundamental differences between sperm and eggs. Sperm are mobile, he noted, while eggs are sedentary; and in this disparity lay the foundations for 'active' masculinity and 'passive' femininity.

Darwin's stereotype of the coy female slotted in with the general mood of the era. A popular ideology gives you the flavour – the Victorian 'cult of true womanhood' claimed to be based on, but also informed by, science. 'True women' were expected to be pious, submissive and interested only in domesticity. They lacked passion and were not interested in sex, even after marriage. Procreation was a marital duty to be performed as part of the sacred oath, but with no relish or enthusiasm.

Darwin's theory of sexual selection chimed with these ideas. But it was perilously controversial nevertheless, far more so than natural selection. As we discovered in the last chapter, female choice was its Achilles heel. Giving such evolutionary agency to the fairer sex stuck in the throat of the Victorian patriarchy and made Darwin's second theory distinctly unpalatable. The concept of sexual selection might have quietly disappeared were it not for a British botanist who, some seventy years later in 1948, provided the experimental data to support Darwin's gender stereotypes and convert them into universal law.

Darwin's authenticator was Angus John Bateman, a young but distinguished plant geneticist working at the John Innes Horticultural Institute in London. Bateman hatched an ambitious plan to legitimize Darwin's 'general law' that males are ardent whereas females are coy. Bateman noted that Darwin had based these gendered roles on observation alone and, without empirical support, the great man had been 'at a loss' to 'explain the sex difference'. Bateman's self-imposed challenge was to offer an empirical lifeline to Darwin's ideas.

To do so, Bateman turned his attention away from plants and on to the minuscule flies that seem to magic out of nowhere to waft around rotting fruit. *Drosophila melanogaster*, the humble fruit fly, may be an annoying pest to most frugivores but it is the geneticist's best friend. These tiny insects reach sexual maturity in a matter of days, lay hundreds of eggs and, most crucially, can be bred in the lab to display obvious genetic mutations. Cultured strains of flies with odd-coloured eyes, no eyes at all or deformed stumpy wings act like visible name tags that make it possible to trace genealogies.

Bateman placed 3–5 adult male flies and the same number of females, all with different physical mutations, in a glass container and let nature take its course – a sort of *Love Island* for mutant flies with grotesque names like 'hairy-wing', 'bristle' and 'microcephalous' (alias the no-eyed tiny head).

After a few days, Bateman inspected the next generation to deduce who had been mating with whom. The parental mutations were all heterozygous, meaning each fly had a dominant mutant gene and a recessive normal one. Basic Mendelian genetics would therefore predict that if, for example, Bristle and No-eyes got it together, one quarter of the flies should bear their father's stiff bristles, one quarter would be blind like their mother, one quarter would be both bristly and blind, and one lucky quarter would be normal with no mutations at all. Using this basic principle Bateman estimated how many offspring each male and female had produced.

Bateman's monstrous *Drosophila* mating party was an elegant, if somewhat macabre, way of studying heredity in the days before paternity testing and decoding genomes. It enabled him to establish who had fertilized whom, without having to watch the flies' mating behaviour. It was an exceptionally elaborate undertaking neverthe-less, since, in all, he conducted no less than sixty-four mating experiments. I know from experience that working with fruit flies is extremely fiddly, not to mention sticky and smelly, work. *Dros-ophila* are just three millimetres long, so inspecting thousands of young adults for bristles or hairy wings must have been a monu-mental challenge even to the most committed pedant.

Bateman pooled his results from all sixty-four experiments and presented them in two basic graphs, which plotted reproductive fit-ness (i.e. number of offspring) against number of matings. The second of these two graphs has now become the stuff of legend, reproduced in millions of zoology textbooks around the world. It features just two lines: one for the male, which thrusts skyward and illustrates that more matings equals more offspring; the other line, for the female, rises limply before levelling off, demonstrating that females gain nothing from mating with more than one or two males.

'Bateman's gradient', as it is widely known, proved that for males competition was such that while some flies were studs, others were duds. Females, on the other hand, showed little difference in their reproductive output. Bateman's results revealed that the most suc-cessful male fruit fly produced nearly three times as many offspring as the most successful female. Whereas a fifth of all males failed to sire any offspring at all, compared to just 4 per cent of females.

This variability in reproductive success cast a dull shadow over the female of the species. It implied that sexual selection acted more strongly on males than females, which were pretty much guaranteed to breed at capacity. So, as well as being branded coy, females were also deemed evolutionarily irrelevant and their behaviour unworthy of scientific scrutiny.

Despite the fact that Bateman had only tested Darwin's theory on fruit flies, he felt confident that his conclusions could be extrapolated to far more complex organisms, like human beings. He proclaimed that a dichotomy in sex roles, namely 'an indiscriminating eagerness' in males and a 'discriminating passivity' in females, was the norm across the animal kingdom. 'Even in a derived monogamous species (e.g. man) this sex difference might be expected to persist as a rule,' Bateman concluded.

These fixed sex roles, Bateman proposed, were underwritten by anisogamy. A female's reproductive success is constrained by her finite number of large, energetically expensive eggs, whereas for males, a limitless supply of cheap sperm means their reproductive output is only curbed by the number of females they can win over and mate.

Despite his positive result, Bateman's empirical lifeline for Darwin's gender typecasts initially joined sexual selection in academic oblivion. The young botanist returned his attention to plants and probably never looked at another *Drosophila* again, unless perhaps to swat one off his fruit bowl.

All was not lost, however. Twenty-four years later a Harvard zoologist by the name of Robert Trivers named Bateman's experiment as the key reference in what proved to be one of the most influential biological papers ever, cited over eleven thousand times. Trivers' classic 1972 essay on 'Parental Investment and Sexual Selection' exhumed Darwin's theory of sexual selection along with Bateman's supporting evidence, gave them an uncritical polish and turned the coy female and the philandering male into one of the guiding principles of evolutionary biology.

Trivers argued that whichever sex invests least in offspring will compete to mate with the sex that invests the most. At the root of this inequality, again, is anisogamy and the perceived need for a female to protect her already substantial egg investment, while the male splurges away indiscriminately.

Trivers' paper landed at the birth of sociobiology (now also known as behavioural ecology, a new field of evolutionary biology that focused on animal behaviour) and became part of its foundation. Textbooks with chapter titles like 'The Reluctant Female and the Ardent Male' became every biology student's bible, with Bateman's gradient prominently featured. So pervasive was this biological law that it seeped out of science and into popular culture, prompting a celebration of scientific theory in some of the most unlikely places.

'It has been said that a man will try to make it with anything that moves – and a woman won't. Now the startling new science of sociobiology tells us why,' crowed *Playboy* magazine in 1979, in an in-depth feature bursting with *Schadenfreude*. 'Darwin and the Double Standard' accused feminists of defying their biological heritage, claiming, 'Recent scientific theory suggests there are innate differences between the sexes and what's right for the gander is wrong for the goose.' The lavish exposé closed with a licence for its readers to shag around, rubber-stamped by science: 'If you get caught fooling around, don't say the Devil made you do it. It's the Devil in your DNA.'

The sentiments celebrated by *Playboy* magazine continue to haunt the science of evolutionary psychology, which seeks to find explanations for human behaviour from evolution. Scientists from Alfred Kinsey to David M. Buss (author of *The Evolution of Desire*) have focused on male promiscuity against the baseline assumption that this behaviour resembles mating strategies in the animal kingdom prescribed by anisogamy. Some even justify the worst human male behaviour – rape, marital infidelity and some forms of domestic abuse – as adaptive traits that evolved because males are born to be promiscuous while females are sexually reluctant.

The trouble with this universal law is it's not universally true. Just ask the lioness. The first cracks in Bateman's paradigm appeared even before Trivers had raised the botanist's fruit fly paper from the dead, but few, it seems, wanted to recognize them. The intellectual weight of Darwin's name, bolstered by Bateman's paradigm and Trivers' rising star, encouraged most zoologists to look the other way if they encountered a philandering female. It took a bunch of birds, both feathered and human, and a toolkit for investigating criminals to set the record straight.

HOORAY FOR CHEATING BIRDS

The red-winged blackbird, *Agelaius phoeniceus*, is the menace of North American farmers. Every spring these glossy onyx songbirds with their fancy scarlet-and-gold epaulettes descend upon America's wheat belt in great numbers and proceed to peck their way through the nation's grain crops. Back in the 1960s the farmers' answer was simple: they shot as many of them as they could. This trigger-happy solution to songbird control didn't sit so well with the federal government's wildlife protection agency. So, in the early 1970s, US Fish and Wildlife Service conservation scientists decided to trial a cunning alternative inspired by their zoological intelligence of the blackbird's sex life.

Male blackbirds are polygamous. Each male defends a territory big enough to maintain a harem of up to eight females. The US government scientists therefore proposed that a more humane, if somewhat unconventional, solution to the annual spring slaughter would be to vasectomize the males, the rationale being that a blackbird that shoots blanks cannot sire further pests. Minus his testicles, Mr Blackbird and his harem could live out happy sterile lives without perpetuating the scourge.

Somehow this idea managed to leave the drawing board and enter a field trial in Colorado in the spring of 1971. A trio of male scientists captured eight male blackbirds on some marshland near Lakewood in Jefferson County and set about tying their tubes – a delicate operation on a bird that weighs little more than a tennis ball. The blackbirds, once they had recovered from the anaesthetic, were returned to their territories where they 'recovered rapidly and without any apparent ill effects'.

The females meanwhile had already started to lay eggs. Over

the next nine days these were all removed from their nests so that the scientists could be sure that any eggs in their experiment were the product of liaisons with the newly sterile males.

Much to the scientists' surprise and dismay 69 per cent of subsequent clutches from the vasectomized males' territories turned out to be fertile. There were three possible explanations for this extraordinary result, the most likely being that the fiddly sterilization process had in fact failed. So all eight males were sacrificed and their gonads carefully scrutinized under the microscope. Compared to unsterilized males, the scientists were reassured to find all had severely reduced testicles and 'cheesy material' in their distended vas deferentia tubes, suggesting the operation had indeed done its intended job.

Still confused, the scientists wondered if perhaps the female blackbirds had been storing sperm they'd received from the males pre-sterilization, which they then used to fertilize their eggs at a later date. But when they examined the insides of thirty females at various stages of nesting they deduced that viable sperm could not have been retained for long enough to have fertilized the eggs.

There was only one remaining explanation. The females must have been having sex with unsterilized males in neighbouring territories. This was an unthinkable scenario. In the 1970s the human sexual revolution may have been in full swing, but female songbirds did not play the field. 'Well over nine tenths [93 per cent] of all passerine subfamilies (songbirds) are normally monogamous,' wrote the ornithologist David Lack with authority in 1968. 'Polyandry* is unknown.'

---

* Let's not get our polys in a twist. Polygamy is probably the most familiar term and is used to describe a pattern of mating in which an animal, of either sex, has more than one mate. Polyandry, on the other hand, is a specific form of polygamy where it is the female that has more than one mate. It is easy to remember by breaking it down: poly- being Greek for 'many' and -andr meaning 'male'. Polygyny is the form of polygamy in which the male has several female partners. Again, a little Greek helps, as -gyne denotes woman. Then there's monandry in which the female has just one mate, but as you'll discover this term doesn't need to get used that much.

The baffled scientists were forced to publish that male steriliza-tion is an unsuitable means of controlling blackbird populations, citing a suspicion that 'female promiscuity' may be to blame. Their sheepish admission represented a fail for pest control but it foretold a revolution in our understanding of female mating behaviour.

Aside from a few polygamous exceptions like the red-winged blackbird, songbirds were long thought to be the very models of monogamy. It's easy to understand why. It is obvious to even the most casual observer that most birds breed in pairs, with the male and female working tirelessly to build a nest and furnish their demanding chicks with the food they need to fledge. This is often done within the grounds of human habitation, making their paral-lel domestic set-up easily observed and even romanticized.

'Be thou like the dunnock – the male and female impeccably faithful to each other,' proclaimed the Reverend Frederick Morris in 1853. Morris was a keen ornithologist and author of a popular Victorian book on birds that encouraged his readers to emulate the modest lifestyle of the 'humble and homely' dunnock, aka the common brown hedge sparrow (*Prunella modularis*). The good reverend was unaware that this was in fact permission for his female flock to seek out a second lover and copulate over two hundred and fifty times with both males in order to start a family. A decree which, as Nick Davies, the Cambridge zoologist who first documented the female dunnock's liberated sex life, wryly noted, would have resulted in 'chaos in the parish'.

It turns out there is a world of difference between social and sex-ual monogamy. Birds do social monogamy very faithfully, with some species even maintaining pair bonds for life; but sexually, it's another story.

According to the avian biologist Tim Birkhead, professor of behavioural ecology at the University of Sheffield, this discovery rocked the world of ornithology. 'It was the biggest discovery in bird biology in fifty years,' he told me on a blustery birdwatching trip to the RSPB's Bempton Cliffs in Yorkshire. It was the height of the breeding season and the white chalk cliffs were alive with thousands of soaring kittiwakes, head-bobbing gannets and growling guillemots. All around us pairs of seabirds were busy doing showy courtship dances, building nests, feeding chicks and secretly cheating on each other.

'Because Darwin said females were monogamous, that's what everyone believed for a hundred years,' he said. 'Even when there were cases that were blatantly non-monogamous, people would say they'd made a mistake or make an excuse like the female had a hormone imbalance. It was swept under the carpet.'

We now know that 90 per cent of all female birds routinely copulate with multiple males and, as a result, a single clutch of eggs can have many fathers. It turns out that the flashier the male, the more likely the female is to be unfaithful. Birkhead recently discovered that species displaying the greatest sexual dimorphism hide the greatest infidelity, the most extreme known case being Australia's superb fairy wren. As the name suggests, males boast some rather splendid blue seasonal plumage and will even woo the female, a classic little brown job, with carefully plucked yellow flowers. The male's flamboyant courtship is rewarded with cuckoldry. At dawn his partner will slip away and have sex with the neighbours. As a result, over three quarters of the chicks in their nest, diligently provisioned by her social partner, have actually been sired by different males.

Female birds are sneaky about their extra-pair affairs. Their duplicity was only rumbled using forensic DNA fingerprinting co-opted by zoologists to test the paternity of their eggs.

The first person to utilize this novel technology to investigate female songbird fidelity was Patricia Gowaty, a virtuoso scientist now in her seventies, and a distinguished former professor of evolutionary biology at UCLA. Gowaty is a feminist firebrand. This inspired piece of detective work was an early breakthrough in a career dedicated to fearlessly questioning what she calls 'the standard model' of sex differences in behaviour. It was also her first taste of being ignored by the male scientific establishment.

'I got a lot of flak from this study,' she told me over the phone in a disarming southern drawl. 'It was terrible, Lucy. It was as though I'd discovered something, but it offended so many people that it was unbelievable.'

Gowaty's subject was the eastern bluebird, *Sialia sialis*, the cobalt-coloured songbird associated with happiness and featured in Disney's 'Zip-A-Dee-Doo-Dah'. A much-loved avian superstar as wholesome and all-American as apple pie, and Gowaty was effectively calling her a Jezebel. It was never going to go down well, but Gowaty was shocked at the depth of prejudice amongst her peers.

'They couldn't imagine that females were anything but benign. They would never seek an extra mate, I mean . . . it's a sin! They didn't say it like that, but that was what was in their mind.'

At a meeting of the American Ornithological Society a well-known male ethology professor voiced his scepticism by telling Gowaty that the bluebirds in her study must have been 'raped'. This, she explained to me, is physically impossible. Male songbirds have no penis. Both sexes have a multi-purpose hole called a cloaca that is used to transfer gametes and waste. In order for fertilization to happen both male and female must evert the middle section of their cloaca so that they touch in what biologists call a 'cloacal kiss'. This must be done while the male is balancing precariously on the female's back, so the female can call a halt to any unwelcome sexual shenanigans simply by flying off.

'Songbirds have no need for a #MeToo movement,' Gowaty told me. 'It is physically impossible to be fertilized without female compliance.'

The ensuing decade saw a flurry of bird paternity studies and a tidal wave of evidence that could no longer be ignored. Yet somehow female birds remained resolutely coy in the eyes of the (male) ornithological establishment. After all, the law according to Darwin-Bateman-Trivers stated that females had nothing to gain from multiple matings, and everything to lose – if her social partner were to catch her, it was believed an adulterous female would run the risk of being deserted or, worse, killed. So despite the impossibility of rape, the prevailing view was that female birds must be the forced victims of the male's biological prerogative to splurge his seed far and wide.* Even ornithologists like Tim Birkhead debated how female birds were 'suffering' forced extra-pair copulations and how they could trick the male into monogamy.

'The more this oddly puritanical idea is examined, the stranger it appears,' wrote Susan Smith, professor of biological sciences at Mount Holyoke College, Massachusetts, whose long-term study of black-capped chickadees was instrumental in turning the tide. Over the course of fourteen breeding seasons she noted that 70 per cent of female dalliances took place shortly after dawn in the territory of a male higher-ranking than the bird's social partner. It looked suspiciously like female birds shacked up with Mr Average were sneaking off to their neighbour, Mr Fabulous, for some superior genes.

Smith's pivotal study was augmented by the work of Canadian ornithologist Bridget Stutchbury, professor of ecology at York University in Toronto. Stutchbury told me over Skype how she initially fell in line with the established females-as-victims story. That was until she started putting radio transmitters on the backs of hooded warblers in the early 1990s.

Stutchbury discovered that females aren't victims at all, but actively advertising their fertility with a special *chip chip* call that lets neighbouring males know a female's up for some extra-pair sex. By tracking the movements of males, Stutchbury documented how their visits to neighbouring territories coincided with the female's fertile window and the broadcasting of her unique *chip chip* come-and-get-it-guys call.

'Females make a lot of noise when they're fertile,' she told me, 'so we figured they must either be very stupid, or willing participants.'

Female hooded warblers would also leave their territory to scope

---

* A small group of birds, such as ducks, have retained a penis, and forced copulations are a feature of the mating system, which we shall discuss in chapter five.

'out the local talent, but only during their fertile period. This they did by listening to the male warblers' warbling. Females paired with meek males that didn't sing that much would leave their home territory the most and seek out sex with a more strident stud. Subsequent DNA testing showed that these vociferous males did indeed sire the most offspring.

Here was scrupulous evidence of female birds owning their sexual destiny and the paternity of their eggs. But Stutchbury's team struggled to get their pioneering paper published. 'We had reviewer after reviewer tell us that we were just point blank wrong,' she told me.

The reviewers of academic papers are anonymous, but given the field was at least 80 per cent male at the time, the sexual identity of the commentators is easy to guess. Especially when you factor in the level of mansplaining involved in their rejections.

'We even had one reviewer say that "we were kind of dumb" – the only reason females are calling like that is because we (the researchers) are on their territory trying to observe them. The female birds are actually calling at us.'

Stutchbury's paper was eventually published in 1997. It joined other studies on blue tits and tree swallows that also revealed so-called monogamous females actively seeking infidelities with sexier males than their diligent chick-feeding social partner. Together they heralded a new dawn. 'In most bird species, it is likely that females control the success of a copulation attempt and of transfer of sperm,' wrote Marion Petrie, professor of behavioural ecology at Newcastle University, somewhat triumphantly in a review of bird paternity studies in 1998. A simple statement, but one that would never have made it into print just ten years previously.

These philandering female songbirds sparked a 'polyandry revolution' that shook the world of behavioural ecology.

Females across the animal kingdom began wresting back control of their sexual destiny and egg paternity from the assumed dominance of males. DNA testing techniques resulted in a cascade of other females – from lizards to snakes to lobsters – having their fidelity revoked. Polyandrous tendencies were discovered in every vertebrate group, and amongst invertebrates polyandry was proclaimed the norm rather than the exception. True till-death-do-us-part sexual monogamy, on the other hand, proved to be extremely rare, found in less than 7 per cent of known species.

'Generations of reproductive biologists assumed females to be sexually monogamous but it is now clear this is wrong,' Tim Birkhead admitted in his 2000 book *Promiscuity*.

The establishment had finally accepted females would actively seek out sex with multiple males. But why they would do so remains a source of controversy. The Bateman-Trivers paradigm predicts that females have nothing to gain from 'excessive' matings, so their lusty advances make little sense to devotees of this 'universal law'.

'The mystery still is, what do females get out of it?' Birkhead told me on our birdwatching trip to Bempton Cliffs.

LICENTIOUS LANGURS

Not everybody finds female promiscuity so bemusing. I made a pilgrimage to rural California to meet one of my academic icons – the renowned American anthropologist Sarah Blaffer Hrdy, professor emerita of University of California, Davis. The six-foot Texan, now in her seventies and still oozing glamour, greeted me with open arms, a specially made pie and a tour of the walnut farm she runs with her husband Dan, a fellow academic. With great pride Hrdy explained how they'd created this green idyll from dust – planting and nurturing native trees and hedgerows to restore the habitat to a more natural state. Which is kind of what she's done in her academic field. Hrdy has spent over forty years weeding out sexist dogma and sowing new theories that allow the true nature of females to flourish. Hrdy was the first person to ever challenge 'the myth of the coy female' and is known by many as the original feminist Darwinist.

'I prefer the term "distaff Darwinist",' she told me. 'I am not sure those applying the term to me define it the same way I do. To me, a feminist is just someone who advocates equal opportunities for *both* sexes. In terms of evolutionary theory, that means someone who considers selection pressures on females *as well as* males.'

As a graduate at Harvard in the early 1970s Hrdy found herself at the epicentre of the new science of sociobiology and in the orbit of its wunderkind, Bob Trivers.

Hrdy was the only female graduate in her class and the focus was firmly on male animals. The halls were heavy with testosterone. 'Sexism was built into the sciences at Harvard back then,' she told me. The textbooks of the time considered female primates only as mothers, fundamentally nurturing and with zero competitive edge. Female primates were 'invariably subordinate to all the adult males' and sexual behaviour was understood to play 'a small part in the life of an adult female'. As such they were 'relatively identical' and considered scientifically dull. There was much weeding to be done.

Hrdy started a project to investigate mysterious reports of infanticidal behaviour amongst male Hanuman langurs (*Presbytis entellus*), elegant Old World monkeys with long grey limbs and charcoal faces native to the Indian subcontinent. But from the very start, it was the females that grabbed her attention. Her first glimpse of a langur was a female near the Great Indian Desert in Rajasthan moving away from her family group, and sashaying up to a band of bachelors to solicit them for sex.

'At the time, I had no context for interpreting behaviour that merely seemed strange and incomprehensible to my Harvard-trained eyes. Only in time did I come to realize that such wandering and such seemingly "wanton" behaviour were recurring events in the lives of langurs.'

Hrdy hit the library, dug deep and discovered her langurs weren't the only 'wanton' female primates out there. Many social species exhibit an aggressive sexuality, bordering on nymphomania, especially when ovulating. A wild chimpanzee female will produce only five or so young in a lifetime yet she will avidly engage in some six thousand or more copulations with dozens of males. When ovulating she might solicit every male in her community and have sex 30–50 times a day. Barbary macaque females are equally lustful, with one female recorded having sex at least once every seventeen minutes with every sexually mature male in the group (of which there were eleven). Savannah baboons have been documented badgering males for sex with such frequency during their oestrus that their lustful advances are even refused.

'I think the Greek word from which our term "oestrus" is derived provides as good a descriptor as any for female proceptivity around the time of ovulation,' Hrdy told me, 'deriving as it does from "a female driven to distraction by gadflies".'

In dozens of female primates, such gadfly madness provokes a frenzy of sexual activity that far exceeds what's required to fertilize the egg on offer. Some have even been observed seeking out sex when there is no egg to fertilize. Hrdy documented her langurs seducing alien males from outside the troop during pregnancy. Whilst others, such as orangutans and marmosets, show continuous receptivity and, like humans, are sexually active throughout their cycle.

Such excessive behaviour is not without risks. Retaliatory attacks by possessive males, venereal disease, increased predation risks from leaving the troop, not to mention all the energy required to power this 'excessive' sexual activity make female philandering anything but cost-free. Far from being monandrous, therefore, females are apparently under strong selection pressure to be promiscuous.

'In retrospect, one really does have to wonder why it was nearly 1980 before promiscuity among females attracted more than cursory theoretical interest,' Hrdy has said.

What's more, many females appear to be rather enjoying it. It may come as a surprise, but all female mammals have a clitoris. For some, like the domestic ewe, it's a rather discreet affair, whilst for others, like the spotted hyena we met in chapter one, it's a flamboyant eight-inch organ that bulges forth like a penis. There is a wealth of morphological diversity in between. What you see is however just the tip of the iceberg. In humans, this richly innervated organ extends ten centimetres internally, with two arms that wrap around the vagina, and is the seat of the female orgasm. Whether other female mammals are able to derive such pleasure from theirs has been the subject of much debate, with a gang of male scientists saying, 'No,' and a load of female scientists screaming, 'Yes!'

The populist British anthropologist Desmond Morris was one of many such men to have an opinion. He pronounced the human female orgasm to be 'unique among primates' – its function being to maintain the monogamous pair bond. The shameless pleasure-seeking of many female primates would indicate otherwise.

For a start, most female primates have been documented masturbating – both in zoos and in the wild. The British prima-tologist Caroline Tutin documented a wild female chimpanzee nicknamed Gremlin displaying 'a fascination with her own genitals' that resulted in 'rubbing objects such as stones and leaves' against them. A party for one that suggests there may have been some pleasure involved. Jane Goodall also noted female chimpanzees fondling their privates 'and laughing softly as they do so' (less *le petit mort* and more *le petit rire*). Female orangutans have been observed flaunting their dexterity by masturbating with the balls of their feet, and tiny tamarins by using their tails or 'soft surfaces' until they enter a 'trancelike' state.

Without the ability to ask, it is of course hard to ascertain exactly how much satisfaction a bonobo is deriving from the homespun French tickler she's fashioned out of twigs, but a few bold scientists have attempted to gauge whether female primates are indeed orgasmic. A close observation of the sexual behaviour of wild stumptail macaques by Suzanne Chevalier-Skolnikoff concluded that females did indeed climax, and even provided a helpful drawing of the monkey's characteristic round-mouthed O-face.

In an experiment that could only have taken place in the 1970s, and must still have caused a few raised eyebrows in the pub afterwards, the Canadian anthropologist Frances Burton attempted to

If you ever wondered what a monkey's O-face looked like, now you know. Suzanne Chevalier-Skolnikoff's research into female orgasm in macaques provided this handy illustration of the classic 'frowning round-mouthed expression' made by both sexes during climax.

bring an end to the debate by manually stimulating three female rhesus macaques to orgasm in a lab using an artificial monkey penis. The monkeys were each strapped into a dog harness and wired up to a heart monitor while Burton valiantly provided each female with her allotted five minutes of genital manipulation.

It's hard to imagine a less sexy, more clinical setting yet each of the monkeys clearly exhibited three of Masters and Johnson's four copulatory phases used to define orgasm.* Two of the monkeys even displayed the 'intensive vaginal spasms' that characterize the

---

* William H. Masters and Virginia E. Johnson were a pair of American sex therapists who pioneered groundbreaking research into the human sexual response and treatment of sexual dysfunctions and disorders, from 1957 to 1990. In 1966, they proposed a four-stage 'linear' model of human sexual response

human female orgasm. Burton tentatively concluded that rhesus females do indeed have the ability to climax. But she noted that under natural circumstances copulations were far briefer – lasting a mere matter of seconds. The level of stimulation required to bring the monkeys to orgasm could only be achieved in the wild after several copulatory bouts with stimulation that was accumulative. Say, for instance, with a succession of males.

To evolutionary psychologists like Donald Symons this orgasmic response is 'dysfunctional' – the result of the clitoris being little more than a useless homologue of the penis, with no adaptive function. According to Symons females didn't actually evolve their own orgasms. Any sexual pleasure derived by the female is simply a happy biological accident, made possible thanks to a shared developmental blueprint with the penis.

'Are we to believe that the clitoris is nothing more than a pudendal equivalent of the intestinal appendix?' wrote Hrdy in *The Woman That Never Evolved*. To her eyes, the variety of clitoral morphology screams adaptation. 'I cannot understand why these old canards persist.'

Comparative anatomical studies are thin on the ground but in promiscuous multi-male breeders – like baboons and chimps – the clitoris is especially well developed, reaching an inch or more in length. It is also positioned at the base of the vagina, where it can receive direct stimulation during sex. Which suggests that these females are being rewarded with significant pleasure for the sex they have with multiple partners. But why?

Hrdy's big idea is that the function of all this non-conceptive sex is to manipulate males.

While studying her langurs in India, Hrdy observed males from outside the group routinely killing unweaned infants, as part of a

---

based on some ten thousand recordings of changes in participants' physiology: (1) excitement, (2) plateau, (3) orgasm and (4) resolution.

troop takeover. This infanticidal behaviour, she realized, is a toxic side effect of sexual selection and male competition for mates. Rather than his having to wait two to three years for the female to wean another male's baby before being available to mate again, by infanticide the new leader forces the bereaved mother into oestrus, making her readily available for fertilization. As a defence against infanticide, Hrdy theorized, females are driven to have sex with invading males. This has the effect of protecting the lives of their babies by confusing paternity. Hrdy's theory explains why that first female langur she saw was heading off from her group to solicit a band of foreign males and why others were seen having sex with alien males during pregnancy. Far from being 'wanton', their overt sexual behaviour is in Hrdy's eyes 'assiduously maternal' – evolution's cunning ploy to increase the survivorship of their young.

Perhaps unsurprisingly, a theory that sees murderous baby-eating males outsmarted by maternally driven sexual hedonism was considered somewhat heretical at first. Hrdy's ideas have been attacked by evolutionary psychologists blinded by Bateman and even the Vatican, which once sent a 'hostile' envoy to a conference she gave on the meaning of sexual intercourse. Others chose to simply belittle the Harvard scientist and her work. Hrdy recalls one male colleague's 'mortifying' response to her theory: 'So, Sarah, put it another way – you're horny, right?'

A wave of supporting evidence has now seen Hrdy's theory of paternity confusion incorporated into mainstream academic thinking. It's not something that sits comfortably with human morality but infanticide by males is now understood to be widespread amongst our primate cousins, being strongly suspected or actually witnessed in some fifty-one species. In almost all cases, males only attack when entering the breeding system from outside it, specifically targeting unweaned infants. The same pattern is true of male lions, which will kill cubs when taking over a pride. Which means the lioness I accidentally seduced was biologically compelled to try to have sex with me, not simply because she fancied the sound of my tinny roar, but so I didn't wind up murdering her cubs.

Other known infanticidal males across the animal kingdom, from dolphins to rats, may also be the reason why their females are wired for promiscuity, but Hrdy is keen not to make blanket generalizations. 'It is important to study females as flexible and opportunistic individuals who confront recurring reproductive dilemmas and trade-offs within a world of shifting options,' she emphasized, making clear the pitfalls of universal paradigms.

There are a host of other likely benefits to female philandering that include a quest for superior genes or a greater chance of genetic or immune system compatibility that then increases the survivorship of offspring. In essence, female promiscuity leads to healthier offspring, as it means that a mother does not have to put all her precious eggs in one basket.

'That said, sexual selection probably explains most cases of infanticide by males, and manipulating information about paternity is one of the few practical options females have. I expect females in quite a few species will have converged upon this solution to a pretty terrible predicament,' Hrdy explained.

Paternity confusion isn't just an insurance policy against infanticide. It also encourages males to care for and protect infants. Some of the best evidence for how mothers benefit from manipulating information about paternity derives from those salacious sparrows we met earlier in the chapter. Dunnock females, as we know, are typically polyandrous, taking two lovers – an alpha and a beta. Both males will help the female with the task of provisioning the chicks. Research has shown that the males actually calibrate the number of mouthfuls of food they bring, in accordance with how often they managed to copulate during the female's fertile period. DNA fingerprinting has revealed that dunnock males were often, but not always, accurate in their paternity 'assessments'.

Hrdy revealed that all across the primate world males are manipulated into caring for babies that may or may not be theirs. This unequivocal fact pours cold water on the common theory that monogamy is a female's best strategy because males will only care for young they know are theirs. This concept may be fashionable with male evolutionary psychologists penning populist bestsellers in urban offices, but it holds no credence with the anthropologists who are actually out in the field observing the wild behaviour of female primates. Hrdy notes how studies of barbary macaques and savannah baboons have all shown libidinous females using sex to draw multiple males into a web of possible paternity, which results in them routinely babysitting, carrying or protecting other males' offspring. Our ancestors may have done the same.

'Non-conceptive sexual behaviour, while not increasing fertilizations, increases infant survivorship, and as such is the ultimate female reproductive strategy,' Hrdy says. She's certain this polyamorous maternal tactic could have been especially useful for our hominid ancestors in helping them raise an exceptionally slow-maturing infant, which demands years of care before reaching independence.

How female sexuality has been transformed in the intervening 4–5 million years is open to speculation. Humans are a socially monogamous species today, but then so is the superb fairy wren. Evolutionary biologists like David M. Buss may relish the idea that all women are ultimately seeking monogamy in order to provide the best support for their kids, but if women were so naturally inclined towards fidelity then why, wonders Hrdy, is their sexuality so culturally controlled? Whether the restraining tool is slanderous language, divorce or, worse, genital mutilation there is a near universal suspicion that women will engage in promiscuous sex if left unchecked. An alternative perspective, endorsed by Hrdy, sees the potency of female sexuality being such that patriarchal social systems evolved in order to curb and confine it. Which makes a woman's degree of faithfulness a highly flexible affair. One that cannot be divined by her fixed gametal destiny, however popular the paradigm, but is dependent on her circumstances, and the various options open to her.

TESTICLES DON'T LIE

If you want to know how promiscuous the female of a species is, there are a couple of big bulging physical clues which are known to provide a reliable metric. The weight of the male's gonads (in relation to his body weight) provides a generalized rule of thumb, or perhaps rule of testicle, that lays bare the female's sexual habits.

Take two common British butterflies. The cabbage white has testes that rival its prodigious appetite for homegrown brassicas, i.e. enormous. The meadow brown, however, from the family Satyridae – named somewhat ironically after the lustful woodland Greek god – has tiny ones in comparison. This physical disparity mirrors the different mating strategies of the females – cabbage whites are polyandrous, whereas meadow browns are not.

The phenomenon was first documented in primates by the Australian zoologist Roger Short, who noticed the surprising variance in the testicle size of great apes, which curiously fails to correspond with the animal's overall mass. Great big silverback gorillas may be the terrifying heavyweights of the great apes, three times as big as a male chimpanzee, but their jewels are less than a quarter of the size – the equivalent of a dainty pair of strawberries to the chimpanzee's large pears.

It all comes down to sperm competition. Large testicles produce sperm at a higher rate, giving the male a better chance of either filling a female's reproductive tracts so that no further sperm can be deposited, or swamping out the male that got there first. A silverback's muscular mass enables him to control access to a harem of females that remain faithful to him. Female chimpanzees on the other hand copulate 500–1,000 times for each pregnancy, with many different males. The physical result of all this philandering is that the chimpanzee sports testicles ten times the size of a gorilla's in relation to body weight, so he can drown out the competition. Human testicles, you will be keen to know, fall somewhere in the middle of these two.

Across the animal kingdom – from butterflies to bats – testis size turns out to be a sure-fire indicator of female fidelity: the bulkier the balls, the faster and looser the female. For many species, such as lemurs, big gonads are a seasonal affair timed to coincide with ovulating females. Once the need to breed is over they deflate like party balloons with a slow leak, sometimes to a tiny fraction of their high-season size. If sperm are so cheap then why the seasonal adjustment? After all, as Dawkins said, 'The word excess has no meaning for a male.'

'History has not been kind to this pronouncement,' Zuleyma Tang-Martínez, emeritus professor of biology at the University of Missouri, has wryly noted.

The other side of the Bateman equation – that males are preternaturally 'eager for any female' and their 'fertility is seldom likely to be limited by sperm production' – has also come under critical fire. A number of scientists have pointed out that the cost of a single spermatozoon may be trivial in comparison to an egg, but so far science has failed to discover a male that delivers just one swimming wonder at a time. Each ejaculate contains millions of sperm along with a cocktail of critical bioactive compounds, the inevitable expense of which racks up the overall biological bill* such that in mammals, for sure, we now know that the combined energetics of a single ejaculate is in fact greater than that of an egg.

As such, semen production is generally limited and 'sperm depletion' a genuine concern, with most males needing time to replenish

---

* Some of these semen proteins provide the female with direct benefits that will actively encourage her to mate with many males. Bush crickets deliver protein-rich 'nuptial gifts' in their semen that provide the female with a handy post-sex snack to feed her developing eggs. Some semen proteins are known to stimulate egg production and even increase life expectancy. Texan field crickets are one of many species with prostaglandins in their semen, which boost the female's immunity. These prostaglandins are found in the semen of a wide range of animals from insects to mammals, suggesting that for a wide range of females having lots of sex with multiple partners is actually good for them. It could explain why the females of those promiscuous species have an increased lifetime fecundity compared to those that don't.

their stocks after a big spend. In humans, for example, complete recovery can take as long as 156 days.

Some, like the spiny lobster or bucktooth parrotfish, deal with this depletion issue in a Scrooge-like manner by budgeting the size of their ejaculate according to the reproductive value of the female. A female's age, health, social rank or previously mated status will dictate how much sperm he is prepared to spend. Others simply refuse a female's sexual advances. An Australian species of stick insect, with little more to do all day than munch leaves and make like a stick, was presented with a fresh female once a week yet only managed to rouse himself to mate 30 per cent of the time. Other species from European starlings to Mormon crickets have all been observed turning down sex on a regular basis. Some, like the male great snipe, have even been known to chase away soliciting females.

BATEMAN REDUX

All of which places males in the females' choosy shoes. In fact, even *Drosophila* fruit flies, the original poster boys for the licentious lifestyle, have been documented acting coy in the face of unrestrained female flies, which undermines Bateman's paradigm in the most fundamental way. The experiment was the handiwork of Patricia Gowaty, who cunningly tested anisogamy theory by using three species of *Drosophila*, all with different size sperm in relation to eggs. These included one species in which males have freakishly large sperm that actually outsize the female's egg. Would their giant gametes restrain their sexual behaviour in comparison to the males with regular tiny sperm?

Unlike Bateman, Gowaty didn't simply infer their mating behaviour from the resulting offspring, since this would not reveal the whole sexual story – only the winning tries. Instead she diligently watched the three-millimetre flies play out their mating game 24/7 in order to get a more nuanced appraisal.

In each of the species she found some females were as active in approaching males (or more so) as males in approaching females, and some males were as discriminating (or more so) as females, even though they differed in their anisogamy. All of which suggests that gamete size had nothing to do with their sexual strategy. 'The labels "choosy, passive females" and "profligate, indiscriminate males" did not capture the variation within and between species in pre-mating behaviour,' Gowaty has said.

Gowaty is by no means the only critic of Bateman's original experiment. Tim Birkhead noted that female *Drosophila melanogaster*, the fruit fly used by Bateman, can store sperm for 3–4 days. This would reduce their need to re-mate within the limited four days of the experiment. Had Bateman chosen a different species of fruit fly, Birkhead mused, which didn't store sperm, the geneticist might have got very different results. Not only that, it turns out *Drosophila melanogaster* semen is spiked with anti-aphrodisiacs that warp female behaviour, making her wait longer before mating again; a chemical chastity belt that induces a certain coyness and also has the potential to have skewed Bateman's results.

The ultimate test of a scientific experiment is, of course, to repeat it. The replicability of experiments is considered an essential part of science. Given the 'foundational nature' of Bateman's seminal paper, Gowaty felt it was 'important to know that Bateman's data are robust, his analyses are correct and his conclusions are justified'.

So Gowaty took it upon herself to redo Bateman's elaborate study using the same methodological protocol and the same mutant flies. This was no mean feat. First her team had to locate the exact same strains of deformed *Drosophila*, then came the even harder part – deciphering Bateman's methodology.

'I think I know Bateman's study better than anyone else on earth,' she announced, somewhat wearily, over the phone. The antiquated paper was 'very hard to understand – a mishmash of stuff'. Gowaty and her detective partner Thierry Hoquet managed to dig up Bateman's original lab notes from some dusty old archives and re-analyse his original data. Gowaty's laser-like scientific mind identified significant issues that pointed to a bad case of confirmation bias. Bateman's methods 'had flaws, variances, statistical pseudo-replication, and selective presentation of data'. Gowaty concluded that 'Bateman's results are unreliable, his conclusions are questionable, and his observed variances are similar to those expected under random mating.'

In short, Gowaty says, 'Bateman's paper is just a travesty.'

For a start, Bateman counted mothers as parents less often than he counted fathers, which is of course a biological impossibility since you need both to make a baby. He failed to recognize that inheriting one of his grisly marker mutations from both parents – say stumpy wings *and* an eyeless micro-head – might be somewhat lethal. When Gowaty repeated his experiments, sure enough, many of the double deformity offspring 'dropped like flies'. These matings would therefore have been invisible to Bateman and resulted in him overestimating subjects with zero mates and underestimating those with more than one.

Bateman's famous discovery that only male reproductive success benefits from promiscuity actually only applied to the last two of his experiments, which included these deathly double mutants. These (now questionable) results were, for no logical scientific reason, combined and given their own graph, which delivered the famous Bateman's gradient seen in millions of textbooks around the world. The first four experiments actually showed that females also benefit from playing the field, albeit to a lesser degree. Gowaty points out that had Bateman pooled all his results into one graph, and analysed the data accordingly, he could have laid claim to the first ever evidence of the benefits of female promiscuity. But Bateman, and everyone after him, focused only on the results that fitted Darwin's proposition of promiscuous males and choosy females.

'Bateman made a result that was consistent with his expectations,' Gowaty told me. 'I mean he is dead now and it's kind of mean to out him, but he did not do a good job on what he set out to do.'

While some errors were buried deep and required replication to be exposed, others, like the biased pooling of results, were 'so obvious' that Gowaty can't understand why the hundreds of scientists citing Bateman didn't also take note. 'The fact that Bob Trivers didn't recognize it seems an extraordinary error,' she told me.

Trivers made Bateman's paper famous, although it's unclear whether the Harvard hotshot had read the botanist's paper that carefully. 'Most females were uninterested in copulating more than once or twice,' Trivers wrote of the mutant flies, which is not something even Bateman could have known without a psychic connection to the female fruit flies since he hadn't actually observed their behaviour. Bateman simply inferred matings by counting the resulting offspring, so his experiment only revealed how many males had successfully fertilized females, not how many males the females had actually had sex with – a key over-simplification that was perpetuated far and wide. More damning still, when Tim Birkhead quizzed Trivers in 2001 about why he had ignored the first graph, which showed female reproductive success did indeed benefit from multiple mating, and focused only on the second, which showed that it didn't, he told Birkhead 'unashamedly that it was pure bias'.

THE SLOW DEATH OF THE CHASTE FEMALE

Paradigms are powerful things, especially those infused with an insidious cultural bias. Their overwhelming influence can dazzle even the most diligent scientists, limiting the way we see the world and obfuscating fresh perspectives from outside the box. For too long, the Bateman world view blinded us to the idea that females were not only soliciting sex with multiple partners, but that this licentious behaviour could be a benefit to them and their offspring. In the dance of sex Bateman's principles assumed females were always led by males and were therefore not worth studying. But it's impossible to understand what one sex is doing without considering the other and by denying females any variance in their reproductive success it meant we misunderstood the strategies of not only females but their male partners too.

According to Zuleyma Tang-Martínez, many scientists continue to question or ignore their own findings because their results do not accord with Bateman. 'It is not unheard of for some journal reviewers and editors to rebuff papers that report increased female reproductive success as a function of number of mates because "Bateman showed in 1948 that such results are not possible".'

Gowaty and her collaborator, Malin Ah-King, also dug up dozens of empirical studies which actually demonstrated females being indiscriminate, or that males had been choosy; yet the authors failed to recognize these results. 'That is very curious to me,' Gowaty said. 'It shows people are afraid.'

Their dominant nature makes paradigms hard to shift. Even those built on quicksand take time to topple. The fact that Bateman's paradigm is not supported by Bateman's data would seem to be something of a terminal setback, empirically speaking. Bateman's gradient not only fails to predict what goes on in 'man', it fails to predict what goes on in lions, langurs or even (in Gowaty's meticulous analysis) in Bateman's own evidence from *Drosophila melanogaster*. While there are some species which do follow Bateman's gradient, there are now dozens of experimental studies in a wide range of animals from prairie dogs to adders that demonstrate females do increase their reproductive fitness from promiscuous behaviour.

To Gowaty, and others, this means Bateman's principle should be considered a hypothesis, as opposed to a fact, and taught as such. But others cling to the few species that do obey Bateman's predictions and maintain that 'evolved sex roles ultimately rest on anisogamy'.

'People just believe in anisogamy theory as though it was handed down by the gods,' Gowaty told me with obvious exasperation. 'We have all been led down some primrose path that we believed in without carefully thinking about what in the hell is going on. There must be something about the powers that be that depend on these profound differences between males and females. I think anisogamy theory somehow reinforces the generalized misogyny in the world.'

The debate around Bateman's work has certainly become political. The paradigm's foundations were constructed in Victorian chauvinism and they've been toppled by feminist scientists. But the F-word has a polarizing effect, so much so it can undermine solid science. Gowaty believes her open politics have prevented her scientific papers from being widely read, especially by those who should be reading them. When Trivers was interviewed a few years ago by Angela Saini for her book *Inferior*, which charts the prevalence of sexism in science, he claimed not to have read Gowaty's 'God-Jesus paper'. At the University of Oxford today, where Bateman's paradigm is still taught, Gowaty's critical studies don't make the reading list, as they are considered to be 'very political perspectives'.

'Empirically minded biologists hear that dreaded F-word and assume it must mean "ideologically driven",' Sarah Blaffer Hrdy said to me. 'What they overlooked of course was how *masculinist* many of their own assumptions were, how androcentric the theoretical foundations of their own Darwinian world view was.'

Was Bateman wrong about everything? Possibly not. Anisogamy might have tilted the evolutionary playing field in certain species, but it does not begin to explain all that's going on with sex roles. Gametes differing in size is just one factor among many affecting costs and benefits of different strategies. Bateman saw sex roles as fixed: choosy and passive for females versus indiscriminate and competitive for males. But the picture that's now emerging is that sex roles are not only more variable, but also more flexible and fluid than previously understood. Social, ecological and environmental factors and even random events all have the power to shape their nature. In many species of cricket, for example, the availability of food can cause sex roles to switch within a lifetime, turning choosy females into competitive crickets and making profligate males discriminate when supplies are low.

Throughout the animal kingdom, females have busted out of the

Playboy mansion and are leading sexually liberated lives for the benefit of themselves and their family, with no shame attached. Darwin's sexual stereotypes may have been psychologically compelling to generations of male scientists, but they've been overthrown by an army of sexually assertive warblers, langurs and fruit flies, and the intellectually assertive females studying them.

The female of the species is emerging from the shadow cast by Bateman's rigid paradigm, and revealing a wealth of sexual strategies that expand, rather than restrain, Darwin's concept of sexual selection. In the next chapter we'll meet some voracious females whose sexual appetites have sucked the romance out of sex and revealed how, rather than being a cooperative affair, conflict is often at its heart.

CHAPTER FOUR

Fifty ways to eat your lover:
the conundrum of sexual cannibalism

Who can fathom the mind of a spider?
Keith McKeown, Australian naturalist (1952)

Seduction is an awkward game for many males. The stakes are high, as is the suitor's vulnerability. Timing, technique and a certain amount of chutzpah are all required to secure success. But when the object of your desire is a ferocious predator that eats animals that look like you for breakfast, finding a mate becomes a dance with death.

This is particularly true in the case of the male golden orb weaver spider (*Nephila pilipes*). The female is Goliath to his David: around 125 times his mass, and armed with giant fangs that deliver a potent venom. To seduce her, the male must gingerly traverse her enormous web – a succession of tripwires designed to sense the slightest vibration – then clamber on board her gargantuan body and copulate, all without triggering her hairpin attacking instinct. His chances of running this sexual gauntlet with life and limb intact are slim at best. For the male golden orb weaver spider sexual disappointment takes the form of a grisly death, his would-be lover literally sucking the life out of him in a matter of minutes, before hurling his desiccated carcass on to the burgeoning heap of failed suitors below.

News of such shocking female behaviour did not escape the attention of Darwin, although his handling of the horror is highly euphemistic. In *The Descent of Man* he details how the male spider is often smaller than the female, 'sometimes to an extraordinary degree', and must be extremely cautious in making his 'advances', since the female often 'carries her coyness to a dangerous pitch', which is, I suppose, one way of putting it.

Darwin's androcentric account eventually dares to spell it out by noting how a fellow zoologist by the name of De Geer saw a male that 'in the midst of his preparatory caresses was seized by the object of his attentions, enveloped by her in a web and then devoured, a sight which, as he adds, filled him with horror and indignation'.

The female spider's penchant for rolling dinner and date into one was an affront to male Victorian zoologists on several counts. Here was a female that deviated from the passive, coy and monogamous template by being vicious, promiscuous and unquestionably dominant. She also represented something of an evolutionary conundrum. If the point of life is to pass on your genes to the next generation, then eating your potential sexual partner instead of mating with him seems maladaptive. Yet sexual cannibalism is common amongst spiders of all kinds, along with a host of other invertebrates, from scorpions to nudibranchs to octopus. The most famous is probably the praying mantis, a femme fatale who devours her lover's decapitated head, while his truncated body continues to valiantly thrust away behind. The existence of such behaviours has prompted generations of zoologists to suppose that evolution itself has lost its head.

'If a male spider didn't have to mate with a female, he would certainly avoid her,' Dave Clarke, the head keeper of invertebrates at ZSL London Zoo, explained to me.

Clarke should know. He's in charge of the walk-through spider exhibition at the zoo, where it's possible to stroll freely amongst

enormous webs and take selfies with the orb weavers at their centre.
I've been many times, but until Clarke showed me around I had no
idea that the big spiders in the middle of the webs are always female.
Male spiders are generally weedy itinerant creatures with little time
for web-building or indeed hunting, and their fangs and venom sacs
are often puny in comparison. The female spider is the one to sport
the more toxic venom and construct the most elaborate webs. These
extraordinary feats of engineering become their domain; a place to
hunt, mate and nest.

As keeper, part of Clarke's job is to breed the animals in his care.
In over thirty-five years at London Zoo Clarke has successfully
bred 'pretty much everything' from giant anteaters to moon jellyfish. To achieve this end he must get to know his subjects intimately.
'There is always a certain amount of voyeurism involved with the
job,' he admitted.

It's Clarke's responsibility to figure out the equivalent of soft
lights and mood music for the animal in question. This is easier
said than done. It's not just giant pandas that are hard to mate in
captivity. Every taxon brings its own complications. But it is the
spiders that generate his greatest performance anxiety.

'It's incredibly intense. When you're talking about spiders, it's
funny to think of being in their mindset, but you can't help it.
You're really feeling for the male. Not just whether he's going to
mate or not, but if he's going to survive,' Clarke explained. 'You're
putting yourself in his position and you almost feel that fatal stab,
when it all goes wrong.'

Some of the most dramatic sexual liaisons Clarke oversees are
between the bird-eating spiders in his care. These are the titans of
the spider world, with legs that can span up to thirty centimetres. I
remember one scuttling under my foot as I walked along the street
in Cairns, Northern Queensland – like a scene out of the eighties
horror movie *The Hand* – and I almost jumped out of my skin. As
their name suggests, these spiders have reversed the traditional
food chain order and regularly consume birds or even rodents that
might otherwise consider one of their more diminutive arachnid

cousins a tasty snack. Breeding such colossal beasts in captivity is a gladiatorial event.

'It's incredibly exciting to watch. You're sort of transfixed, probably because you can see it on such a big scale. The males have hooks on their front pair of legs and they have to use those to hold up the females' fangs during mating so that she can't bite him. It also means he's in the best possible position to reach forward with his pedipalps and insert them,' Clarke explained.

Male spiders don't have a penis. Sperm is transferred to the female using a pair of leg-like appendages, known as pedipalps, that sit either side of the head. These aren't connected to his testes, however, so a male spider must first squirt semen from his abdomen on to a special 'sperm web', which he then siphons up, like a water pistol, and stores in a large bulb at the end of his pedipalps. He is now locked and loaded and ready to approach the female, with caution.

Position is everything with spider sex, and every species has its preferred angle from the spider sutra.* Most bird-eaters favour face to face, although one bold Brazilian species will flip a female back into missionary position for easy access. The male must reach underneath the female's abdomen to insert his pedipalps into the female's twin genital slits, one at a time. In the case of the

---

* This actually exists. Between 1911 and 1933, the German Ulrich Gerhardt compiled an unequalled amount of data on the reproductive behaviour of 151 spider species. It's fair to say he was obsessed with spider sex. This obsession started at school and by the time he was an adult he had documented sex in spider species from 102 genera and thirty-eight families. His precise documentation included not only the position favoured by the species, but also how many individual thrusts a spider made with each pedipalp. There was fine detail on insertion behaviour – be it 'groping' or 'hammering' – along with the male's success rate. Gerhardt paid particular attention to 'flubs' – failed insertion attempts. These he counted and exposed for each species. Thanks to Gerhardt we know that male spiders are major flubbers. In twenty species flubs were observed to be 'usual' and occur 'often'. Given the pressure the male spider is under, a little stage fright is perhaps to be expected, but frequent flubbing is surely yet another evolutionary oddity in the spider.

bird-eating spiders, all this must be negotiated whilst holding on to the female's fangs.

'I remember a mating with a redknee bird-eater [*Brachypelma hamorii*] where we only had the one female and one male. Just as he was getting into position, she stuck her fang right through the top of his body. That was it, she literally pinned him to the ground with a centimetre-long fang and there was nothing we could do about it,' Clarke confessed.

Clarke told me he is always on standby with a pot or ruler, poised to intervene if things start to get ugly. Damaged suitors can be rescued and reused, even with lost limbs, especially if they are in short supply. But once the fang is in, venom and digestive enzymes are swiftly injected and the male's organs melt into a spider Slurpee, ready to be sucked up by the female.

'I took on a certain amount of personal failure for that male,' Clarke added, with obvious remorse. 'It is an artificial setting and it's me that's throwing him into the lion's den. So you're thinking, what did I do wrong? Because he's obviously trying to do his best.'

When it comes to seducing big predatory females, Clarke has picked up a few tricks over the years, the most crucial being to ensure the female has been wined and dined before introducing his eight-legged Casanova. 'Hunger is the main reason she'll go for him. If the female hasn't eaten for a while her first thoughts are going to be eating. The male's first thoughts are going to be mating, because that's what he's here for.'

Reproduction may be the point of life, but male and female spiders are working to different timeframes. Females are not just bigger, they also tend to live several times longer than the males. The female redknee, for example, has been known to live for up to thirty years, whilst males are lucky to survive to ten. This introduces a certain amount of conflict to their union. Females want to spend time fattening themselves up in order to lay lots of healthy eggs, so they're in no hurry to mate. At the start of the breeding season, or when they're young, females are likely to have food, and not sex, on their mind. The male on the other hand only ever has one objective, and that's to locate a female and mate as soon as possible.

A male spider's interests lie in passing on his genes, so he wants to mate, but also secure paternity. For the female spider, like many of the females we met in the previous chapter, monogamy is not always such a good idea: she wants offspring with the best possible genes, so she can either be choosy about who she eventually mates with or mate with multiple males, increasing the odds that some of her spiderlings will win the genetic lottery.

'With spiders, the female is really in control of reproduction. Not the male,' Clarke told me. 'The females live much longer. They can also store sperm; in some species for up to two years. So, if they accidentally eat a male, there's plenty more where he came from. They can wait.'

In Darwin's time reproduction was assumed to be a harmonious affair with both sexes cooperating to create the next generation. This romantic notion seems rather quaint today. In the last few decades we've begun to appreciate how females and males across the animal kingdom frequently have incompatible sexual agendas. Love is a battlefield and sexual conflict is now understood to be a major evolutionary force that works antagonistically between the sexes. The tug of war of opposing interests provokes an evolutionary arms race of adaptations and counter-adaptations as each sex tries to outfox the other and get what they want.

Take the langurs we met in the last chapter. An invasive dominant male wants to start siring offspring with fresh females as soon as possible, so he kills their babies to rush them into oestrus. But the females have evolved their promiscuous sexual strategy as counter-attack.

Nowhere is this sexual conflict more extreme than among spiders. The possibility of cannibalism places the ultimate selection pressure on the male to evolve creative solutions to counter the lethal threat of a hungry female.

At the most basic level, many orb weaver males have learned to wait patiently at the edge of the female's web until their paramour is consuming her lunch, possibly one of their love rivals, before making their move. Others, like the male black widow, can actually smell if their fancy is hungry from the sex pheromones on her silk, and keep a wide berth if she is. Then there are those that arrive on their date lugging dead things wrapped in silk – the spider equivalent of a box of fancy chocolates – to occupy the female's jaws while the male gets down to business with his palps.

So far, so sensible. But evolution didn't stop at the male spider simply monitoring the female's digestive state. Sexual conflict has endowed many males with more devious manoeuvres, and as a result furnished spiders with the kind of sex lives that would make even Christian Grey blush.

The nursery web spider, *Pisaurina mira*, is one of around thirty arachnids known to engage in a little light bondage during sex. The male sneaks on to the female's web and ties her up, using a pair of specially evolved extra-long legs so he can keep clear of her fangs while looping his own silk threads around her limbs. With the female restrained, the male can mate safely and at a leisurely pace, taking time to insert his pedipalps multiple times, increasing the chance of sperm transfer and fertilization. Once the job's done, the female releases herself from her silken fetters while the male beats a hasty retreat.

Darwin's bark spider, *Caerostris darwini*, has upped the ante by introducing oral sex into the mix. He binds his lover first with silk and then salivates on her genitals before, during and after copulation. This sexual behaviour has not been witnessed outside of mammals and although its function in spiders is enigmatic, it might serve to digest the sperm of any previous suitors, giving the genital drooler an additional paternity advantage.

Threesomes are the safest way to go for male dotted wolf spiders, *Rabidosa punctulata*. Cruising bachelors that happen upon a copulating couple will chance their luck and join the party. With the female already occupied with a successful male, the interloper is less likely to end up as dinner. The author of a recent study observed some genital sparring between the males but on the whole the spider ménage à trois was a surprisingly orderly affair with males politely taking turns to insert their palps.

And then there's the sensible, if impossible-sounding, strategy of 'remote copulation' in which the male escapes with his life, but waves goodbye to his genitals. When danger threatens, the male nephilid orb weaver spider, *Nephilengys malabarensis*, snaps off his pedipalp and makes good his escape, leaving the palp to pump sperm without him. As an added bonus, his mutilated genital plugs the female's epigynum, preventing her mating with another male. The downside, however, is that the self-inflicted eunuchs are themselves functionally sterile – a one-hit wonder.

The prize for creepiest anti-cannibalism stratagem goes to the wasp spider, *Argiope bruennichi*, so named because of its familiar yellow-and-black striped body. The most successful males seek out an underage female and then guard her until she's on the cusp of sexual maturity. When she's shedding her exoskeleton for the final time before adulthood, her unhardened body is vulnerable; she cannot move, let alone attack a male. It is at that moment the male makes his move and mates with her. It's a successful strategy: copulations with moulting females resulted in 97 per cent male survival compared with only 20 per cent in conventional sex with a hardened adult female.

A FATE WORSE THAN DEATH

Evolution has worked hard to protect the male spider against the threat of pre-copulatory death by female fang. But is sexual cannibalism always just an awkward after-effect of the female's dominant size and unrestrained appetite? Many biologists have argued that to be the case over the years, the loudest of those voices belonging to the Harvard evolutionary biologist Stephen Jay Gould. The US 'evolutionist laureate' claimed in one of his popular 'just-so stories' for the American *Natural History* magazine that sexual cannibalism could never be beneficial. He even went so far as to question whether the behaviour was indeed common enough to warrant close inspection.

'If it occurred always, or even often, and if the male clearly stopped and just let it happen, then I would be satisfied that this reasonable phenomenon exists,' he wrote in 1985.

To Gould the rarity of this apparently aberrant behaviour was evidence for his belief that not every trait we see in nature has been finely tuned by selective forces to ultimately increase an animal's survival and reproductive success. Some are simply accidental by-products of other adaptations, in this case the 'indiscriminate rapacity' and magnitude of the female.

Gould was a brilliant writer and revolutionary theorist but he had never stepped inside the shoes of a spider that's trying to find a mate. Clarke, however, has. For him, there is one thing that's even worse than his intrepid suitor getting fanged, and that's being ignored.

'There's been a lot of spider sex over the years. And the really bad sex is where the female is just not receptive. There's so many cases where you put the male in, you've got all psyched up. You've got the music on and the lights turned down low, and the female is just not moving. She's paying no attention whatsoever. That is really frustrating, and that happens a lot.'

In the wild males may have to travel treacherously long distances, avoiding a host of hungry predators and fighting off other males, in order to locate a female. They may only have one shot at sex, so getting *noticed* is key, even if it results in death.

Clarke told me about one particularly high-profile occasion when he was trying to mate a pair of British fen raft spiders (*Dolomedes plantarius*). This semi-aquatic species, with a leg span roughly the width of your palm, hunts insects, tadpoles and occasionally small fish from the water's surface. It's a big velvet-brown beauty and one of the UK's grandest and rarest spiders, only found in a handful of wetland sites in the whole of the country. ZSL are trying to boost their dwindling numbers by breeding them in captivity and reintroducing the spiderlings into the wild, which is where Clarke and his finely honed spider seduction skills come in.

Captive breeding doesn't get much more pressurized than when it involves an endangered species, let alone a cannibalistic spider. Clarke had to deal with an audience of anxious conservationists scrutinizing his technique as well as the future of the species resting on his shoulders.

'We are trying to maintain a small gene pool over a long period of time, and it all boils down to those few moments of reproduction. So if that goes wrong, you're stuffed,' he told me.

Clarke did his best to replicate the spider's wetland home with a large tank complete with mini water-weed islands, just like the spiders like it. As well as ensuring that the female was, of course, well nourished before introducing the male, he also made sure the couple had plenty of space. Fen raft spiders have a protracted courtship in the wild, involving the male following the female around for some time, testing the water as it were. So Clarke made sure the female had plenty of room to manoeuvre and would not get trapped in a corner and freak out, which can also trigger an attack.

According to Clarke things were going 'swimmingly'. The male had made a number of tentative approaches by bobbing his body and vibrating his legs in delicate arcs on the water around her. The female appeared tranquil and receptive as the male moved in closer. Once within reach, tactile caresses are understood to pacify an aggressive female and form a key part of a male spider's seduction routine. The fen raft suitor began to vibrate his legs across the female's body and cautiously manoeuvre himself into the correct position to insert his pedipalps. Then, in a split second, she grabbed him and killed him.

'I knew straight away that was it for him. And I felt a terrible sense of regret,' Clarke admitted.

The assembled humans could only look on in horror as the female spider greedily demolished what could have been the saviour of her species. Clarke, no stranger to this kind of all-consuming sexual disappointment, pragmatically saved the female, without

any grudges, for use another day. A month or so later, he was surprised to find a large silken cup in her enclosure. It was an egg sac.
Clarke assumed it must be an immature aberration, but then a further month later the sac suddenly swelled and around three hundred
tiny fen raft spiderlings scampered out. It seems male fen raft spiders can be fiendishly quick on the draw with their sperm pistol
pedipalps, even when getting munched.

'Even in that microsecond of her grabbing him he'd managed to
insert his palp and pass on sperm. And it is just incredible that that
could happen,' Clarke told me, wide-eyed. 'I felt elated. I absolutely
did not believe in any way that she'd actually mated with him.'

The fen raft male may have died a grim death, but he successfully
fertilized the female's eggs nevertheless. So his life, although short,
achieved its purpose. What's more, sucking dry her lover's body
may well have nourished the female's eggs, giving her spiderlings a
better chance in life. The male's sacrificial act benefits both the spiderlings' mother and their (now deceased) father and could be
considered an act of extreme paternal care.

GOOD VIBRATIONS

Most people don't think of the word flamboyant when describing
a spider. The majority of arachnids are creepy little brown jobs,
their drab appearance providing necessary crypsis when hunting or
avoiding detection by sharp-eyed predators. The male peacock
spider, a type of jumping spider from the *Maratus* genus, flouts
this rule quite spectacularly. He is the Liberace of the arachnid
world – an outrageous performer who, just like his avian name-
sake, employs an extraordinary iridescent tail-fan to win his mate.

When approaching a female in the scrub of his native Australia,
this fuzzy little four-millimetre wonder stages an unexpectedly
elaborate dance routine by abruptly lifting his furry abdomen into
a vertical position and unfurling two shimmering flaps decorated

with graphic blues, oranges and reds that could have been designed by Gianni Versace. This peacock arachnid waggles his gaudy butt-fan, whilst bobbing his body up and down, stomping his feet and waving a pair of oversized legs in the air. This exuberant routine, part Fred Astaire and part Village People, can go on for up to an hour until he's close enough to make his move.

It is an undeniably charming spectacle, made all the more endearing by the fact that the peacock male is, of course, dancing for his life. Up to three quarters of peacock suitors are terminally dispatched by an unimpressed female. The unique combination of pathos and panache has turned this tiny antipodean spider into an unlikely internet star. Videos of his kaleidoscopic display, set to tunes like the Bee Gees' 'Staying Alive', have garnered millions of views on YouTube.

'I love them with all my heart,' Damian Elias, associate professor at University of California, Berkeley, told me from behind a pair of thick-framed glasses and a mop of unruly hair. I had already guessed his feelings by the plethora of toy spider kitsch decorating his lab alongside posters of indie rock bands I'd never heard of. Elias has been studying spider seduction for almost two decades (although he doesn't look old enough) and has combined his twin passions – spiders and music – by discovering that the peacock male isn't just dancing, he's dropping major beats too.

Scientists had long believed that female peacock spiders judged their suitors on looks alone. We humans are a highly visual species, so it's easy for us to fall into that assumption. But arachnids inhabit a very different sensory world to us. Most spiders, despite having eight eyes, can't see very well, if at all. Much of their hunting is done by detecting surface vibrations which are imperceptible to us, using specialized slit-like organs on their legs. These vibrations are often amplified by a web, which is essentially an extension of their sensory system. The peacock spider is unusual in that it has evolved acute eyesight as well, in order to stalk and pounce on its prey, namely bugs and other spiders.

The peacock spider's two outsized marble-like eyes have both a telephoto lens and colour vision – an extraordinary evolutionary achievement for an ancient invertebrate – so they're well suited to detect the male's outlandish fan dance. But Elias has discovered the female peacock spider is also using a seismic sense that's imperceptible to us.

'I've always had a keen interest in how animals perceive the world,' Elias told me; and the more alien that world, the better.

Elias has gained entry to the spider's secret sensory realm using an unlikely array of high- and low-tech kit that allows him to act, see and hear like a spider. With great pride he showed me his laser Doppler vibrometer – a $500,000 machine that uses laser technology to measure infinitesimal surface vibrations. It was developed in the 1960s and is commonly used by industrial engineers to inspect jet aircraft safety. It's also been co-opted by spies looking to eavesdrop on indoor conversations from outside a building. Osama bin Laden's fate was sealed when the CIA detected his voice vibrating on the windows of a compound in Pakistan.

Elias uses his vibrometer to spy on the earth-moving character of the male peacock spider's pulsating dance moves. The vibrometer is also hooked up to a speaker, which allows Elias to translate these minute substrate oscillations into 'songs' – sound waves that are audible to our ears.*

To record accurately the 'songs' required getting the right stage for the male spider to perform on. In the wild, peacock spiders leap around among leaf litter, rocks and sand. Elias needed an even substrate for the laser technology to work. Graph paper and tin foil proved too stiff, ringing like tuning forks with confusing air currents and ambient noise. Through trial and error, Elias designed his

---

* Seismic communication may be a new discovery for humans, but it is surprisingly widespread across the animal kingdom. Elephants use their toes to sense the trumpets and stomps of faraway friends. Golden moles target bite-sized termites by feeling the bugs' footsteps. Some South American frogs thump the ground with their bulging air sacs to greet mates and rivals. And for most invertebrates, substrate-borne vibrations are more commonplace than airborne audio.

ideal spider disco dancefloor: a pair of American Tan nylon tights, minus their cloth gusset, stretched taut over a tapestry hoop.

When I asked Elias where he sourced his stockings, he sheepishly admitted, 'It's kind of embarrassing actually. I steal them off my mother-in-law.'

Peacock spiders move faster than our eyes can perceive. So Elias had hooked his pilfered pantyhose dancefloor up to a high-speed macro camera that captured the male's calisthenic moves on a grand scale and in slow motion, so the spider's visuals can be matched with his beats.

To complete the experimental set-up we needed a female peacock spider that could be controlled, which is where the carbon dioxide gas came in.

'Sometimes you have to do bad things to get results,' Elias told me with palpable regret as he euthanized a female for me to manipulate. The minute spider, no bigger than a grain of rice, needed to be fixed on to the end of a pin with hot wax – a fiddly task that involved staring down a microscope while wielding a roasting soldering iron. A steady hand was needed, which I lacked. By the time my temptress was secure she was hot in all the wrong ways – minus a few limbs and reeking of burnt hair. Elias told me not to worry; with peacock spiders, apparently, it's all about the eyes. As long as the two big ones were intact, the male should still perform. If not then Elias had back-ups in what he referred to as 'the graveyard' – a pin cushion hosting half a dozen of his 'corpse brides'.

My singed seductress was attached to a dial at the centre of the dancefloor, so I could turn her and deliver flirtatious looks to the male with one hand. My other hand was occupied with a paintbrush to try to keep the male from escaping. Jumping spiders are capable of leaping fifty times their height without the use of muscles. This is all done using hydraulics, much like how a digger's arm moves: liquid is pumped into their hollow limbs and this extends them. The first three males demonstrated this skill quite dramatically by pinging clean off the pantyhose and into oblivion. Elias remained cool, assuring me they would turn up somewhere eventually (hopefully not in my luggage back in London), and in the meantime 'there are plenty of dead bugs in the lab for them to eat'.

Trying to manoeuvre the female and corral the male was about as easy as patting your head and rubbing your stomach at the same time. My main issue was getting the male to notice my female puppet. Although a peacock spider's eyesight is acute, the only way their tiny brains can process that level of visual acuity is to view just part of the picture. 'It's like looking through a pair of binoculars,' Elias explained. Both male and female need to lock eyes in order to actually see each other.

Eventually, Casanova number four clocked my spider seductress and the game was suddenly on. This was *very* obvious. The male abruptly flung his flashy elongated third legs in the air in a highly theatrical wave and began what's best described as a jazz hands routine. There followed a few seconds of intense stomping and then the reverberations began. There was a deafening buzzing, as if a big booming bee had entered the lab. The male was pulsating his abdomen at a speed of two hundred vibrations a second and the 'noise' it made, once translated from pantyhose tremors to sound waves by the vibrometer and speaker, was deafening. There was no way a female could ignore this male's presence, especially with him furiously waving his technicolour butt-flag at the same time.

There followed a frenzy of ostentatious flag and leg waving, perfectly coordinated to a variety of thunderous beats to which Elias had given evocative names like 'rumble-rumps' and 'grind-revs'. After around thirty intense seconds of this spider flamenco the male had crept close enough to make his final move. As he clambered aboard to claim his prize Elias moved in with his paintbrush.

'That's the end of your fun,' Elias quipped, and brushed him aside so fast my spider suitor had no chance to realize his girlfriend was, in fact, dead. It felt rather brutal after such an extravagant effort, but I reminded myself that, had she been alive, my singing and dancing Casanova could easily have been the lifeless one.

Elias has discovered that the seismic songs produced by the male have around twenty elements and are as complex as any made by humans. Each spider is essentially a freestyling jazz artist riffing on a set formula to make it his own. It was an unquestionably impressive performance, but what did it all mean?

It is Elias' belief that the peacock spider's flamboyant routine is the ultimate dance with death. Its extravagant nature is designed to show off the male's vigour, but also to grab the attention of a female who may be busy hunting (things that look like him) and not exactly thinking about sex.

Creatures with small brains, like spiders, have a limited sensory world, so punching through the environmental noise to get noticed isn't easy. So they employ the same tactic as the blue bling satin bowerbird we met in chapter two and exploit existing sensory pathways to get noticed. But rather than gathering objects that look like their fancy's favourite food to tempt her, many take things a dramatic step further. In other words, a spider suitor will act like lunch to get laid.

Female jumping spiders have their predatory instinct triggered by flickering motions made by prey in their peripheral vision, detected using their smaller eyes. This could be why the male peacock starts his seduction routine with jazz hands as this imitates the jerky movements of a tasty insect. Vibrations are probably also important, especially for other less-visual spiders. Many males start their seduction routine with tremors that mimic a struggling insect in a web. These spark the female's hunting instinct from a distance, before she even sees or smells him.

That combination of visual and vibratory stimuli means the male is effectively stepping into the lion's den dressed as a steak and screaming, 'Eat me!' It's a sure-fire way to get noticed, but this reckless strategy needs to be followed pretty sharply with one that stops the female from acting on her predatory impulse, otherwise the courtship's going to be very short.

'The fact is that spiders are predators. And so one of the best ways to draw attention is to basically trigger those things that are going to make a predator turn around and look for food. But then very quickly after that you have to say, actually, I'm not a fly or a cricket,' Elias explained.

The baroque courtship routine signals that a male isn't lunch after all, but a member of the opposite sex, of the same species and looking to mate. Elias believes this energetic display also indicates the health and vigour of a suitor, enabling the female to decide which males get lucky and which get chomped. Females like big males and it's thought that a male's size is betrayed by his vibrations.

All that fan waving and butt shimmying may also serve another function: to hypnotize the female into quiescence. Shudder-like vibrations, for example, are known to delay a female's attacking instinct, even when there is prey on her web. So these quivers can be co-opted by the male into hip-shaking moves that tap into very basal aspects of neural control in the female.

Spiders may even be able to sense airborne vibratory signals. Like their arthropod cousins the flies, they're covered in long filiform hairs that can pick up air movement down to one ten-billionth of a metre, roughly the width of an atom. These hairs are the reason why it's so hard to swat a fly: they can detect the air particle displacement that swooshes ahead of your hand and make an escape before it gets close. So the peacock male's jazz hands routine may not be a visual signal – instead he's attempting to waft the female into submission. Like other spider species he could also be adding chemical stimulation – be it aphrodisiacs or narcotics – into the mix as part of this sensory overkill.

For the peacock female sexual cannibalism makes total adaptationist sense. It's a win-win: a means of weeding out the weak suitors, so they stop bothering her with a display that could attract the attention of unwanted predators. And she gets a free meal in the process.

Digging into the peacock female's discernment, Elias found they became aggressive when males made less vibratory effort, failed to coordinate their song and dance routine or, worse still, if they didn't pay attention to *her* signals. Just as we saw with the sage grouse and bowerbird, peacock spider courtship is a two-way communication. Albeit with significantly harsher penalties for those males that fail to listen and respond. A peacock female that wiggles her abdomen is not only unwilling to mate, she'll likely eat her suitor instead.

Sexual cannibalism may have seemed perplexing to the blinkered vision of Victorian male biologists, but when viewed from the female's standpoint there are clear evolutionary benefits to eating your lover before, during or after sex. A mother wants the best genes for her babies and needs to be big and healthy to care for them. She has the physical advantage over her bite-sized suitors, so why not eat the duds?

Female spiders are surprisingly devoted mothers. Many carry around their eggs in a special silken sack and, once hatched, they go on to fiercely protect and feed their spiderlings, sometimes with their very own body. Matriphagy is a thing in spiders. The hatchlings of the desert spider *Stegodyphus lineatus* rely solely upon their mother to provide them with food and nutrients, which she does by regurgitating her own liquified innards. Eventually the hatchlings become so greedy they start tucking straight into her abdomen. Within 2–3 hours they suck her dry until all that remains is her empty exoskeleton.

Darwin assumed the exceptional size difference in spiders to be a result of sexual selection acting on the male: a dwarf can more easily escape the 'ferocity of the female by gliding about and playing hide and seek over her body and along her gigantic limbs'. Modern research suggests that, at least amongst the orb weavers, it is more likely to be *natural* selection acting on the *female*, magnifying her body in the interest of maternity: a big momma is better equipped to withstand the excessive burdens of spider motherhood.

'I think spiders are the best organism to study evolutionary biology,' Eileen Hebets told me. 'Historically there has been this very one-sided view of sexual selection and typically it's focused on the male behaviour and whether they get a mate or not. But in spiders, so much of whether they get a mate or not is determined by the female. It forces a much broader view, where you really have to look at the dialogue between the sexes in order to understand what's going on.'

The Charles Bessey professor of biological sciences at the

University of Nebraska is perhaps a little biased. Hebets has devoted her career to unravelling the evolutionary conundrum of spider sex. Recently she stumbled upon a case of sexual cannibalism so preposterously nonsensical, it left even her scratching her head: wilful sexual suicide by the male dark fishing spider.

DYING FOR SEX

On the surface of things, *Dolomedes tenebrosus* is an unremarkable arachnid: a classic little brown job from North America that lives on trees and hunts other creepy little brown things that live on trees. The dark fishing spider's sex life, however, is truly bizarre, even by arachnid standards. Following the insertion of just one of his two sperm pistols, the male becomes cataleptic, curls up and dies. The female has nothing to do with his death. The male automatically expires every single time, and is left dangling by his genitals from the female for up to fifteen minutes, before she eventually bothers to pluck him off and eat him.

'Not only can he not mate with another female, but he has a whole other palp full of sperm that is just wasted. That to me is just an incredible evolutionary puzzle,' Hebets said to me. 'How on earth did that evolve? It seems so maladaptive for these males to obligately die all the time.'

Biologists call this a 'terminal investment strategy', which sounds more like a dull financial plan for the elderly than something as scandalous as copulatory suicide. This perplexing behaviour has in fact been documented in a handful of other arachnids, including the infamous Australian redback, *Latrodectus hasselti*, a suicidal spider which goes to athletic lengths to secure his sexual death wish.

Redbacks are mostly famous for their enthusiastic bite and penchant for hanging out under toilet seats, a cruel marriage that fuels international *Schadenfreude* with headlines like 'Spider Bites Australian Man On Penis Again'.

The redback's kinky reputation was further enhanced when it was discovered that during sex the male deviates from the orthodox copulatory posture favoured by all his fellow *Latrodectus* species – namely 'Gerhardt's position number 3', aka the standard missionary for spiders – and performs an athletic headstand 'with legs flailing', followed by an elaborate 180-degree somersault so that his abdomen lands on the fangs of his date. The female celebrates the arrival of his succulent body by immediately vomiting digestive juices on to the tiny acrobat. She then proceeds to gobble up her lover's rear end, pausing only to spit out 'small blobs of white substances', while he continues to inseminate her with his head-based pedipalp.

This cannibalistic spider-sixty-nine continues for up to thirty minutes, until the male has spent his pedipalp. At which point he excuses himself from the situation and retreats to nurse his not inconsiderable wounds. Around ten minutes later, undeterred by his heavily masticated and partially digested abdomen, he returns to the scene of the crime to insert pedipalp number two and repeat the whole athletic performance. This time the sex is terminal. As he withdraws his second pedipalp, the female enshrines what's left of her lover in silk, to finish him off at leisure.

The benefits for the redback mum-to-be are somewhat more evident in this scenario than they are for the male. But Dr Maydianne Andrade of the University of Toronto has proved that, for male redback spiders at least, male complicity is no evolutionary accident.

Redback females are, like most spiders, promiscuous. Sperm competition means that mating does not guarantee fertilization. So for a male, dying after sex may be no different to dying *before* sex, if the female goes on to mate again. In the case of the redback, cannibalized males receive two paternity advantages. First, it seems they copulate for longer, which leads to their fertilizing more eggs than males that survived. Second, females were more likely to reject subsequent suitors after consuming their first mate. They were well and truly sated. Given that 80 per cent of redback males will never locate a female and die virgins, sexual cannibalism is adaptive since it significantly increases the chance that their one shot at sex hits the target and delivers the desired genetic rewards.

So how did this hallmark case of male complicity help Eileen Hebets explain the dark fishing spider's copulatory suicide? Well, it didn't. Hebets discovered that around 50 per cent of the time the female dark fishing spider re-mates, often multiple times. So the same paternity guarantee did not apply.

Perhaps, like the fen raft spider we met earlier, the dark fishing spider's death was an act of extreme parental care. The 'male as meal' theory has been a popular adaptive explanation for sexual cannibalism, but concrete supporting evidence has been elusive. Stephen Jay Gould was one of several doubters to note that the male's tiny size – sometimes as little as 1 or 2 per cent of the female – would render his intake meaningless. Like feeding a pea to a hungry elephant.

Undeterred, Hebets and her post-doc Steven Schwartz devised a cunning experiment in which some dark fishing spider females were prevented from cannibalizing the male, some were allowed to eat the male and some had the male switched out and replaced at the last minute with a cricket of the same size. The results of their study were conclusive: females that ate their lover had bigger offspring with greater survivorship. What's more, the cannibalistic spider mums did significantly better than those that ate a cricket, suggesting it's not just about the calories; there must be something uniquely nutritious about eating the male.

'There's a lot of data to suggest that if a species cannibalizes a member of its own kind those nutrients are already tailored to that particular species. And so there could be huge advantages to cannibalizing a conspecific,' Hebets explained to me. But despite this evidence, she's still not satisfied that's the whole story.

'If the female can re-mate that's not enough, I don't feel like I've totally solved the puzzle,' she confessed.

When I tell people I'm writing this book, the most common response I get is, 'Are you going to include the praying mantis?'

Humans have a long-standing fascination with the female sexual cannibal that dates back to those sailor-eating sirens of Greek mythology, and probably even further. She's the ultimate femme fatale. A deviant diva whose voracious sexual appetite and erotic dominance both titillate and terrify – perverting the 'natural order' of male supremacy and sexual potency.

This cultural enthralment and its associated stereotypical baggage seeped readily into science. A recent inventory of the language used in scientific papers documenting the phenomenon complained of the recurrent use of 'highly loaded' language promoting 'a negative stereotype of sexually aggressive females'.

Since the time of Darwin the praying mantis and her fellow date-munching arachnids have aroused the interest of scientists, often male, seduced by a female killer that apparently defies the laws of evolution. The true story that's emerging of the sexual cannibal is far more complex, and far less erotic. Sexual cannibalism disguises a multitude of phenomena, none of them sins. Just as the promiscuous female lion, langur and dunnock are not being wanton, but actually assiduously maternal and simply trying to do the best for their kids, so the female sexual cannibal is similarly merely protecting her future offspring's best interests.

The costs and benefits of this enigmatic behaviour vary depending on whether the consumption happens before, during or after sex, but nevertheless sexual cannibalism has been shown to benefit one or even both of the sexes. It likely evolved independently numerous times in different taxa for different reasons, and is maintained by a cocktail of selective forces: as if sexual conflict, sexual selection and natural selection got drunk together and had a very messy night. The result may seem bewildering, but if you untangle the jumbled-up silken threads, then it all starts to make sense. Although the tiny male golden orb weaver, tiptoeing to his death across a giant female's web, may still struggle to agree.

CHAPTER FIVE

Love is a battlefield: genital warfare

In his 1952 opus on the opossum, the zoologist Carl G. Hartman recounts a long-standing belief surrounding his study subject's mode of reproduction: 'Opossums copulate through the nose,' he tells us. According to legend, the resulting nasal babies don't develop in the opossum's slender snout but are instead removed by a timely sneeze; 'after a period of time . . . the tiny fetuses are blown into the pouch'.

North America's only marsupial is indeed a curious beast. Opossums are amongst the animal kingdom's most accomplished actors; their flair for playing dead has evolved into a multi-sensory performance that includes not only lying stone stiff for hours on end and frothing at the mouth but also emitting a green anal slime that positively reeks of death. In reality opossums don't die easily – they are immune to pit viper venom, allowing them to dine on otherwise deadly snakes, and they also appear to be impervious to botulism and rabies. Their bodies are as extraordinary as their behaviour – they have opposable big toes that act like thumbs, crowded mouths that contain no less than fifty teeth, and the female sports thirteen nipples in her pouch to suckle thirteen undercooked babies each the size of a bee.

The list goes on. Opossums do not, however, have sex through their nose. Early naturalists were blindsided by the opossum's penis, which has the unlikely appearance of a fleshy two-pronged fork. They looked for a pair of holes to receive this twin-branched tool, and settled on the animal's nostrils as the logical point of entry. Had anyone bothered to look inside a female opossum they would have discovered an equally bizarre bifurcated system involving two ovaries, two uteri, two cervices and two vaginas. What makes this embarrassment of riches even more extravagant is the arrival of a temporary third vagina which emerges for the sole purpose of giving birth and disappears shortly afterwards like a secret door.

The diversity of sexual anatomy displayed across the animal kingdom is quite astonishing and goes far beyond what is required to simply transfer sperm to eggs. The opossum may have three vaginas but the elephant shrew has none – the female's womb opens directly on to the outside world. Meanwhile, the male elephant shrew sports a penis that's half the length of its body and bursts forth from his belly in the shape of a Z.

Such variation has long been a boon for taxonomists, for whom a close inspection of an animal's genitalia is often the only way to tell apart closely related species that are otherwise physically identical. These descriptions are invariably androcentric. The widespread use of penis morphology in taxonomy means that there is more known about the male genitalia of many (perhaps even most) animal species than about any other aspect of their anatomy, behaviour or physiology. So standard is the practice of genital identification amongst entomologists that an army of insects, such as *Cacoxenus pachyphallus*, the 'thick dick' fly, have found themselves named after their privates for easy ID.

This law of genital diversity is taxonomically widespread; close species of bumblebee, bat, snake, shark and even primate can be readily distinguished by their sex tools alone. The greatest difference between humans and our closest relative the chimpanzee, for example, is not the size of our forebrains, the arrangement of our teeth or even the flexibility of our digits. It is our genitals. The chimpanzee penis has no glans or foreskin, is held firm by a bone (known as a baculum) and as a final flourish its surface is sprinkled with hundreds of small spines. The human penis is a humdrum fleshy tube by comparison: thick, blunt, boneless and (thankfully) spineless.

No body part evolves as fast as genitalia. This implies they must be under some powerful selection forces. But for centuries the science of studying genitals was not considered a seemly focal point for scientific enquiry. It was fine for the lower regions to be catalogued by taxonomists as part of their perfunctory inventory of physical life, but nobody probed the folds for evolutionary explanations as to why such wanton creativity had arisen in the first place.

Darwin is partly to blame. In *The Descent of Man, and Selection in Relation to Sex*, he insisted that the creative exuberance of sexual selection *didn't* act on genitalia. He considered sex organs to be *primary* sexual characteristics – survival essentials and therefore under the utilitarian guidance of natural selection alone. Sexual selection only acted on *secondary* sexual characteristics – unessential frivolities such as bright plumage or unwieldy antlers; the sexual dimorphisms involved in either male–male competition or female choice.

As a consequence there was no need for pudendum to rear its head in the pages of his book on sexual selection. This must have pleased his daughter Henrietta, who edited his work and, if her opinion on phallic-shaped fungi is anything to go by, readily wielded her red pen when faced with anything too racy. In later life, this Victorian matriarch was said to have spearheaded a campaign to rid the English countryside of the obscenely shaped stinkhorn mushroom – *Phallus impudicus* – because of the effect that seeing it might have on female sensibilities. The last thing civil society needed was to read about the graphic ins and outs of animal genitalia.

For Darwin and his followers, when it came to the study of evolution, genitals were left firmly out in the cold. Which is unfortunate as the bewildering smorgasbord of sexual paraphernalia has much to reveal about the tangle of elaborate forces, far beyond the banalities of 'survival of the fittest', that drive evolution to create the giddy forms that so preoccupied Darwin in *The Descent of Man*.

Salvation arrived, a full century later, in the form of a micro-scopic penis shaped like a brush. The year was 1979 and Jonathan Waage, an entomologist from Brown University, quietly published his succinct observations on the sperm-scooping, as opposed to delivering, talents of the damselfly penis. Brown demonstrated that the rows of stiff backwards-facing hairs at the tip of the phallus enabled the male to spring-clean the female's reproductive tract, removing any lingering sperm left by rival males that might com-pete with its own.*

This one tiny multi-purpose penis sparked a revolution. Darwin had supposed that male competition finished once a male had won the female. But Waage's discovery showed that their sperm contin-ued to compete long after a male had 'won' his mate. This put genitals on the frontline of sexual selection and worthy of closer scrutiny. Suddenly the race to fathom all this phallic diversity became 'one of evolutionary biology's greatest enigmas'.

There followed a gold rush of theorizing as creative as the penis tips in question. Male penises had evolved to promote the separ-ation of species by acting as a unique key for the female lock, thereby making hybridization impossible. Or to combat other males by clasping on to the female to prolong sex, giving their sperm more chance of fertilization. Penile complexity indicated the fitness of the owner – the bigger the better – or how many para-sites he might be carrying. Perhaps they even functioned as homegrown fleshy French ticklers to stimulate the female into

---

* Sperm removal has also been posited as the driver of penis shape in humans. It has been suggested that the glans of the penis might be shaped to move previously deposited sperm away from the cervix during thrusting. Consistent with the view of the human penis as a 'semen displacement device', two surveys of college students showed that sexual intercourse often involved 'deeper and more vigorous penile thrusting' following periods of separation or in response to allegations of female infidelity.

releasing her eggs. Investigators engaged in heated debate about which selective forces – sperm competition, female choice or sexual conflict – were the primary drivers of all this creative exuberance.

THE CASE OF THE MISSING VAGINAS

There was one thing missing from this golden era of genital reflection. Science had amassed a robust literature on penile variation, with intricate drawings and exhaustive descriptions, some dating back over a century. Yet when it came to the female counterpart there was almost nothing to be found. This gaping hole in genital research was little cause for concern, however. The general opinion held that female genitalia were little more than simple tubes that received ejaculate – passive and immutable with no cause to influence evolution, much like their owners.

'There was this expectation that females were not so variable, that there was nothing interesting going on there,' Dr Patricia Brennan told me.

Without evidence to suggest otherwise, this data gap became a self-fulfilling prophecy – female genitals were all the same because there was no information to suggest anything otherwise.

That was until Brennan joined the genital examination party and became the first person to ask, what about the vaginas? Brennan has made it her academic mission to document the female receptacle for all this thrusting diversity and use it to solve evolution's greatest enigmas. Described as 'scientifically unstoppable' by the esteemed evolutionary ornithologist Richard O. Prum, her post-doc supervisor at Yale, Brennan's results have transformed scientific thinking and rehabilitated the female from passive victim to active agent of her own evolutionary destiny.

Brennan is now assistant professor of evolutionary biology at the University of Massachusetts, with her lab based at Mount Holyoke College, one of the most venerable female colleges in the

USA. I visited on a drizzly day at the tail end of October. It was break time and bands of young women scurried between handsome redbrick buildings, antique by US standards. Brennan, keen that I didn't get wet, met me in the carpark with a giant umbrella and a beaming smile. The petite Colombian expat's Bogotan roots still linger in her voice and relaxed style. A row of Halloween pumpkins lined up outside her lab hinted at her impish sense of humour. Brennan had instructed her students to fashion them, not with faces but with an assortment of animal vaginas for other students to identify.

Graffitiing fruit with female genitalia is just one way Brennan has given the vagina a much-needed makeover, transforming the female form from something hidden, unspoken and even shameful and restoring it to rightful scientific prominence. Brennan wields the V-word with easy candour and credits a general discomfort with sex as contributing to the data gap. A discomfort she clearly doesn't share.

'In science we all have biases. But I am a woman and I have a vagina, so I wonder what does the vagina look like?' she told me with characteristic honesty.

Like countless others before her, Brennan's genital curiosity was initially aroused by the male member. It was the early noughties and she was in the Costa Rican rainforest doing her PhD fieldwork on tinamous – ancient jungle fowl that resemble a giant grey chicken with a tiny head. Brennan happened to catch this famously shy creature in flagrante and was shocked to witness a brutal union, with the male apparently forcing himself on the female. When the pair disengaged she noticed what looked like a wine-opener dangling from the male's rear end. At first she thought it was a parasite. Then she noticed the male retracting the curly-wurly worm and wondered if perhaps it could be his penis.

'I didn't even know that birds had penises,' she told me.

This wasn't a case of misplaced naivety on behalf of the budding Cornell ornithologist. Most birds don't have penises. As we've already discovered with the songbirds in chapter three, avian sex generally takes place using a multifunctional unisex hole called a cloaca. The male and female evert and momentarily touch cloacas in what's known as a 'cloacal kiss' – a term which loses its charm somewhat once you know that 'cloaca' is derived from the Latin word for 'sewer'.

The 'cloacal kiss' has always struck me as a rather primitive arrangement compared to the more familiar penetrative system, but amongst birds it is actually more recently evolved. The brotherhood of birds that buck the trend by having a secret penis stored in the entrance to their cloaca, which unfurls only during sex, represent just 3 per cent of all bird species. Joining the tinamous in this exclusive avian penis club are emus, ostriches, ducks, geese and swans. Their commonality is that they're all from ancient avian lines. The dinosaurian ancestors of the birds are believed to have used a similar penetrative system,* but some sixty-six to seventy million years ago the Neoaves, which includes over 95 per cent of the world's birds, somehow lost their penis.

This seems careless at best, but evolution must have its motives. Some have suggested it was for reasons of hygiene – prodding around in a sewer isn't great for keeping clear of STDs (but many reptiles happily do it with their own penis/cloaca arrangement). Others have reasoned that penis loss was a weight-saving device for flight (but tell that to the bats who manage to fly around just fine despite a cumbersome penis that's one of the biggest out there in proportion to body size).

Brennan was underwhelmed by existing explanations and decided to find out for herself why progress had removed the modern bird's penis. So, she switched timid and rare wild tinamous for farmed ducks and got her scalpel out.

'When I dissected my first male duck and saw the penis up close,

---

* At the 2018 TetZoo conference I asked Albert Chen – a specialist in fossilized ancestral birds – what dinosaur penises would have looked like. His wide-eyed one-word answer: 'Scary.' For the curious (and brave), googling pictures of ostrich penises will illuminate Chen's sentiment and suggest that the most terrifying thing about T-Rex may not have been his teeth.

I was blown away because it was so huge and so weird,' Brennan told me. She's not exaggerating. Duck penises are amongst the longest in the vertebrate kingdom proportional to body length. The Guinness World Record holder is the diminutive Argentinian lake duck (*Oxyura vittata*) whose penis, when fully erect, is 42.5 cm long – a full 10 cm longer than its modest body. It also corkscrews counter-clockwise and has a base covered in tiny spines.

The oddness doesn't stop there. The penis of the duck, like the antlers of a stag, is a seasonal event. For most of the year a duck's penis shrinks to a tenth of its size and only during the breeding season does it grow, in some species almost exponentially. When not in use it is tucked away discreetly like an inverted sock at the entrance of the drake's cloaca. When the drake is poised to penetrate he pumps lymphatic fluid into his member, and the penis explodes out of his cloaca at 75 mph, unfurling itself in a third of a second like some kind of sinewy party hooter.

Such a decadent appendage doesn't evolve by accident. The popular opinion was that such extravagance was a result of sperm competition between males. In the majority of duck species sex ratios are skewed towards males, so females have plenty to choose from and competition amongst drakes is fierce. As a result duck sex comes in two forms: elaborately romantic or shockingly violent. Males can win females by courting them using baroque displays involving ornamental feathers and funky acoustics sculpted to an aesthetic extreme by female preference. These displays can start months before the breeding season, giving the female plenty of time to select the father of her ducklings. Once she does, she will invite her chosen male to mate by raising her tail in a distinctive solicitation display.

Those males left partnerless take a darker path to parenthood by engaging in forced copulations.* In many duck species lone males

---

* Since the 1970s, the term 'forced copulation', and not 'rape', has been used by most biologists to describe sexually coercive acts in animals. This is in recognition of the fact that rape amongst humans is a much more complicated phenomenon that can occur for complex psychological, social and cultural

band together to ambush and force themselves, en masse, on defenceless females. I once had a visit to my local park blighted by the sight of a female mallard being brutally attacked by five drakes. The female was desperately trying to escape as the males chased after her, grabbing her by her neck and pushing her to the ground. Feathers flew as one after the other they clambered on board to copulate. All the while, she screeched and struggled in a brave yet pitiful fight. It was horrific to watch and it's no surprise that females are often injured, sometimes fatally as they resist their aggressors in these violent scenes.

Amongst mallards 40 per cent of copulations are forced. In such competitive situations the theory goes that the longer the penis, the more chance a drake has of getting his sperm closer to the egg and winning the race. Which means that in this particular battle of the sexes, the female duck is an abused loser. Not only is she the victim of assault by a ballistic weapon but, more importantly, she's robbed of her sexual autonomy. The female duck no longer gets to choose which male fertilizes her precious eggs, which is the ultimate evolutionary blow.

Female mallards are not alone in being prey to coercive sex. Across the animal kingdom males have evolved myriad ways to win and control paternity, whether the female is consensual or not. Insects called water striders (various *Gerris*) have hooks which latch on to the female and prevent her from escaping copulation. In the newt species *Notophthalmus viridescens*, the males massage sneaky hormonal secretions on to the skin of the females they are courting, which act as aphrodisiacs. And then there are the bed bugs (*Cimicidae*), which practise something known as traumatic

---

reasons that do not apply to the likes of ducks and bed bugs. This is a very important distinction to make. One that has eluded a handful of male evolutionary psychologists who have suggested that human rape is biologically determined by Darwinism. Such claims have been met with widespread criticism. The danger in suggesting a rapist lives inside all human males thanks to sexual selection has forced a strict avoidance of the human term when talking about animals.

insemination. Basically, the male has a hypodermic needle for a penis and uses it to jab away at the female's abdomen, forcibly injecting his sperm directly into her body.

The mallard joins this unenviable confederacy of female underdogs, for whom evolution appears to have dealt a duff deck. That was, until Brennan sliced one open.

'When I dissected my first female duck I almost fell off my chair,' she told me. The textbooks told Brennan that duck vaginas were little more than simple tubes, but Brennan discovered the female's reproductive tract is just as complex as the male's. It is not only long and littered with blind pockets but it also spirals clockwise – in the opposite direction to the male's penis.

'I just couldn't believe it. I even thought that maybe there was something wrong with this female. Perhaps she has a disease, to have such a weird vagina.' So Brennan dissected a second, and discovered exactly the same.

'It was such an obvious structure, these pouches and spirals are very big,' she told me. Like the male's, Brennan discovered the female's convoluted structure is also seasonal. Which explains why the only textbook description of a duck vagina described it as a boring tube – the female was dissected outside of the breeding season.

Something else struck her. Brennan knew from previous studies that although over a third of duck matings are forced, only 2–5 per cent of ducklings arise from these invasive copulations. Brennan had a hunch that the female's helical piping, with its opposing spirals and strange cul-de-sacs, might have evolved to thwart fertilization from hostile drakes by obstructing the path of the penis – a homegrown cock-blocker. The male's penis would have lengthened as an evolutionary response to this vaginal steeplechase, the result of an escalating arms race between the female and male played out over many millennia, genital for genital.

Brennan decided to put her theory to the test. She travelled to Alaska and collected sixteen species of waterfowl during the summer breeding season. The species with the longest penises did indeed correlate with those where the female piping presented more of a twisty-turny obstacle course, and forced copulations were rife. In monogamous territorial species like swans and Canada geese the male's penis was a much more modest affair and the female's vagina correspondingly simple. It was obvious to Brennan that the male and female genitals must have co-evolved antagonistically.

'By the end I could predict what the genitals were going to look like before I even opened them up, and that was really cool,' she told me.

The unstoppable Brennan wanted further evidence of the mechanics of the female's penis-thwarting piping. So, with some effort, she managed to persuade a local duck farm to let her test her theory. The drakes at the farm had been trained to ejaculate into a small bottle so that their sperm could be harvested for artificial insemination. Their party trick would enable Brennan to ascertain just how obstructive the female's convoluted vagina really is when faced with an explosive male penis.

Brennan arrived at the farm with a bag full of prosthetic duck vaginas – from basic tubes to the flamboyant spiralling casts of actual ducks, some made of silicone and some of glass. Male ducks were allowed to mate with a female but at the last minute Brennan switched out the female and replaced her with a fake vagina. Silicone casts could not withstand the explosion of the male's penis and shattered. But glass vaginas withstood the force and did indeed prove Brennan's point – the opposing spirals of the female's vagina significantly slowed or even stopped the expansion of the duck penis when compared to a straight tube. When the spiralling casts were used, the duck penises failed to become fully erect 80 per cent of the time, often in quite an inglorious fashion. They either got bottled in a hairpin bend or in some cases unfurled backwards towards the entrance of the vagina.

Brennan proposes that the female can actually select which drake fertilizes her egg by permitting the passage of his penis further into her oviduct. In non-forced situations, drakes will woo the female with their pre-copulatory dance. If the female wants to mate she adopts her receptive posture, lying flat in the water and lifting her tail.

'She does cloacal winking – the universal sign for take me I'm yours,' Brennan explained. When a female duck lays an egg, she's moving a sizeable object and so she has the ability to expand the lumen of her vagina to accommodate it.

'I think this is what is going on during the forced copulations; the female is non-receptive so she doesn't wink and her vagina stays in this crazy convoluted state.' When she is receptive, however, the female duck opens the lumen of the vagina so her mate can get further along her reproductive tract than unwanted males. She may not be able to choose whom she mates with but she can control the paternity of her eggs, which is, of course, the ultimate goal.

'If you look at female ducks having sex it just looks horrible,' Brennan said to me. 'These forced copulations are so nasty and the female seems so helpless – they are small and can't fight the males off. But it turns out there are other ways to fight them off that are subtler and the males can't do anything about. Even though males are forcing themselves on her they're unlikely to get paternity. Her chosen mate is going to get the paternity. Females have the last word. How cool is that?'

Brennan has rescripted this particular battle of the sexes and recast the winner as female. Her work demonstrates how you can't judge a book by its cover; the duck's hidden reproductive anatomy reveals a very different story to the one suggested by their outward behaviour. Female ducks are not passive victims but active agents driving their own evolution, along with that of males.

Such antagonistic co-evolution is of course a conversation, or perhaps an argument, between the male and the female that plays out over deep time. And the only way to understand it is by paying attention to *both* sides of the story.

'There is a lot of serendipity in science. You can't find answers if you aren't asking the questions,' Brennan told me. 'I think you needed to have a woman to look at this to ask the right questions.'

As for the mystery of the modern bird's missing penis, Brennan's female perspective offers a different explanation to the traditional phallocentric weight-saving or disease-fighting ones. She suspects that the loss of the Neoaves' penis is the product of female choice. Females have chosen less coercive males with smaller penises and, over millions of years of this selective bias, the penis eventually disappeared. The penis-free system is undeniably awkward for the male – it is all but impossible to fertilize the female without her consent. A male can mount the female but he struggles to force his sperm inside her. So she retains control over her eggs without having to run the risk of a damaging fight.

It could be that this new-found female power has even led to further significant changes in the male bird's behaviour. Many Neoaves species are socially monogamous, with males sharing the load when it comes to parental care. Perhaps the female's expanded sexual autonomy advanced her conflict with males over parental care also. Choosing a mate that helps around the nest rather than one that doesn't might incite males to compete with one another to provide the best care. With twice the parenting, offspring can hatch earlier and females can have bigger and more frequent clutches, giving the penis-free Neoaves – the most successful of all the bird lineages – their evolutionary edge.

A COCKEYED VIEW OF EVOLUTION

Since her big duck breakthrough Brennan has acquired a lab full of students to probe the overlooked genital equipment of dozens of other female animals. 'Ducks were the gateway drug, but there is much more work to be done,' Brennan told me.

There certainly is. In 2014 the evolutionary biologist and gender researcher Malin Ah-King surveyed the academic literature on genital evolution dating back twenty-five years and discovered that 49 per cent of studies still only examined males' bits, compared to 8 per cent that just looked at females. Less than half the studies saw fit to study both. This bias was unrelated to the sex of the investigator – females were just as penis-focused as men. And rather than getting better, the prejudice appears to have actually got worse since the year 2000.

Ah-King concluded that age-old assumptions about the dominance of males and a lack of variance in females had cast a lingering shadow over the field, despite studies by the likes of Brennan proving otherwise. 'Too often, the female is assumed to be an invariant container within which all this presumed scooping, hooking, and plunging occurs,' she wrote.

Take the earwig, *Euborellia plebeja*. Males combat the female's promiscuity with highly specialized genital equipment. They have not one but two penises, known as virgae, even though the female has just one genital opening. The second virga is a spare in case the first one snaps off. This might seem overly cautious, but genital loss is apparently quite common with earwigs, owing to the unwieldy nature of their penises. In 2005 Dr Yoshitaka Kamimura, the world expert in earwig sex, discovered that the male's virgae are exceptionally long – as long as the male's entire body – and sport a brush-like tip. Much like the damselfly we met earlier, Kamimura supposed the male used his lengthy virga like a chimney sweep's brush to dislodge any of his competitor's sperm before replacing it with his own.

Almost ten years later, Kamimura got around to examining the female's spermatheca (the receptacle for sperm in many insects) and uncovered a very different story. The female earwig has storage organs for sperm, which are even longer than the male's virga. So the male can sweep as much as he likes, but he's only ever going to displace a limited amount of his competitors' sperm; the female retains control of paternity. 'Thus, females seem to beat males,' Kamimura was later forced to admit. It could be the same story with the damselfly, if anyone ever gets around to looking.

This kind of behind-closed-doors, post-copulatory power of paternity is known as *cryptic* female choice and was championed by the world's leading genital aficionado, William Eberhard of the Smithsonian Tropical Research Institute. Eberhard, who just happens to be married to Mary Jane West-Eberhard (one of the coterie of feminist Darwinists I met at Sarah Hrdy's farm), railed against the 'inadvertent machismo' of his field, which saw genital research 'influenced by male-centred outlooks'. Sperm competition, in particular, was generally portrayed as a male-only sport. Typically portrayed as an epic 'race' which pits sperm against each other like Olympian athletes – only the strongest and fastest will win the prize egg. Females were deemed to have no influence over this contest, as if a cellular version of the 100-metre sprint played out inside their reproductive track while they placidly went about their business, powerless to affect its outcome.

In his pioneering book *Female Control* (1996), Eberhard presented the case that female genitalia – be they vaginas, cloacae or spermathecae – are far more than just inert tubes for ejaculate. They are active organs that can store, sort and reject sperm through their architecture, physiology or chemistry. Females can dump the semen of unappealing suitors, actively speed chosen sperm on a fast track to the ova, or let them languish in a tortuous maze of ducts. The way Eberhard saw it, once insemination has taken place, it is the female who sets 'the rules of the game'.

Eberhard's book was groundbreaking. Yet even this advocate of female sexual autonomy stated that vaginas tend to be more uniform, whereas penises are diverse and species-specific. Brennan agrees that females may vary less than males – their anatomy is constrained by the need to perform other practical functions like laying eggs and having babies – but they are still very much worth studying. 'I love Eberhard's book,' she told me, 'but it gave the impression that female genitalia are not worth studying. That males are where all the action is.'

Brennan's ambition is to create the world's first physical library of animal vaginas to catalogue the taxonomic diversity of shape and function. She's already on her way. Her lab was littered with dozens of ziplock bags containing brightly coloured silicone replicas of an assortment of animal genitalia – llamas, snakes, dogfish, ducks and dolphins – like some kind of highly specialist sex shop.

'So many vaginas, such a short life,' she sighed, surveying the rainbow genital mountain on her desk. Brennan only studies anatomy from already dead animals, hence the apparent randomness of her study subjects. She opened a ziplock and handed me a bright-purple bottlenose dolphin vagina. It boasted a large bulbous chamber at the entrance, which narrowed into a plethora of skinny convoluted folds, leading to another smaller bulb that joined the cervix.

Previously people had presumed the folds in the dolphin's vagina evolved to protect the uterus from the harmful effect of saltwater, which can be lethal to sperm. But Brennan has developed another theory. She handed me another cetacean vagina – a harbour porpoise this time. It was like a longer, stretched-out version of the bottlenose dolphin, but instead of folds it had what appeared to be spirals.

'Exactly!' Brennan exclaimed. 'Convergent evolution with the duck! It's nuts! We were barely able to believe our eyes. Dolphins are essentially like ducks in their genitalia.'

This is especially interesting, as dolphins have another key thing in common with ducks: forced copulations.

'They are the masters of sexual harassment,' Brennan told me. Contrary to their cute image, dolphins' liberal attitude to sex has led to them being dubbed 'aquatic bonobos', with reference to the apes we'll meet in chapter eight that use sex in a wide variety of social situations, and not just conception. Not all of this sexual activity is consensual – coalitions of male dolphins will aggressively herd and harass females for sex.* Brennan and her team supposed that the convoluted nature of dolphin vaginas could

---

* Dolphin sexual aggression isn't restricted to other dolphins. There have been numerous reports of innocent victims from other species, most notably humans. In a scene reminiscent of *Jaws*, the mayor of a French seaside village on the Bay of Brest was forced to impose a ban on beach swimming during August high season when a sexually frustrated dolphin named Zafar began sexually harassing a series of beachgoers.

provide the female with a similarly cryptic means of controlling paternity in forced situations as she found in ducks.

Testing this theory in dolphins proved even more challenging. But the ever-resourceful Brennan and her co-investigator Dara Orbach developed an ingenious way of raising their cetacean sex organs from the dead and getting them to perform in the lab. Their 'Frankenstein sex' involved pumping saline from a pressurized beer keg into the penises to feign an erection and then fixing them with formaldehyde so they didn't lose their engorged shape. Brennan and Orbach then inserted these stiff penises into the corresponding vaginas, sewed them together, soaked them in iodine and pumped away in a CT scanner. This enabled them to peer through the tissues and observe the penis–vagina mechanics of dolphin sex hidden inside.

Human vaginas are known to change shape during sex, so recreating the mechanics of copulation with fixed structures was far from perfect. But Brennan said it was good enough to demonstrate how both the bottlenose dolphin and harbour porpoise penises become obstructed by the female's labyrinthine pipes. That is, unless she is penetrated from a very specific angle. Since dolphins have sex in a three-dimensional space there is ample opportunity for the female to adjust her body, even very slightly, so that an unwanted suitor gets sent up a blind alley.

As with the ducks, Brennan's theory rewrites the fortunes of the female dolphin, transforming her from victim to victor in the battle of the sexes. The more we discover about female sexual anatomy, physiology and behaviour, the more centuries of perceived male dominance begin to wane. Females have evolved creative ways to control the insemination of their eggs, even when males are more powerful, numerous or forceful. Recent work has shown that when male eastern mosquitofish evolve longer genitals (known as gonopodia) to harass females with, females grow bigger brains in order to outwit their aggressors.

'Females can retain control anatomically, behaviourally or even chemically. Sometimes it is subtle and sometimes it is not. And these strategies can pile on top of each other. I don't know if it is a paradigm shift, but we are certainly recognizing the biological reality that these reproductive interactions are really complex. There is no reason to assume that if ducks have complicated vaginas, they don't also have chemical ways of thwarting sperm. They probably do,' Brennan told me.

Brennan had further positive news about bottlenose dolphin sex. She believes that females also derive pleasure from it.

'Do you want to see a dolphin clitoris?' Brennan asked me, with a certain amount of glee. Before I could say yes, she dived under a dusty lab bench to prise open the biggest Tupperware box I'd ever seen, filled to the brim with pickled pudenda. A sickly waft of sweet alcohol hit my nose as her arm delved into this formalin soup to retrieve a large lump of fixed flesh with a groove down the centre – like a pair of oversized fleshy burger buns. The genital chunk was big and slippery, and kept skidding out of Brennan's rubber-gloved hands as she attempted to prise open 'the buns' and expose the exterior of the dolphin's clitoris concealed within. When she finally succeeded I got a shock. The dolphin's clitoris had an unnervingly familiar hooded appearance to it. If it wasn't for its generous size, it could have belonged to a human.

As we discovered in chapter three, the clitoris evolved for sexual pleasure and there is huge variety amongst mammals, suggesting there are strong evolutionary forces at play. But compared to the penis, we know very little about its morphology or histology. Brennan is changing that. She showed me cross-sectional slices that she'd made in her lab using a deli counter meat slicer. They revealed the considerable extent of the dolphin clitoral tissues internally. 'There is so much erectile tissue – it has to have a function,' she said to me.

PULLING BACK THE HOOD

Assessing that function has been hampered by a sea of misinformation and cultural baggage. The clitoris is perhaps the only organ even more understudied than the vagina. It arrived on the anatomical map in the mid-sixteenth century, having been 'discovered', somewhat unconventionally, by an Italian Catholic priest called Gabriele Falloppio (1523–62).* When Falloppio's finding was shared, however, with the great physician Vesalius, the founder of modern human anatomy, it was swiftly dismissed. Vesalius proclaimed that 'this new and useless part' didn't exist in 'healthy' women, and was only to be found in hermaphrodites.

This inglorious misunderstanding set the stage for the next four hundred and fifty years, which saw the clitoris routinely lost, rediscovered and subsequently dismissed by the medical patriarchy. Despite complex internal drawings by the German anatomist Kobelt detailing the full extent of the organ in the mid-nineteenth century, the clitoris was characteristically *difficult to find* in modern anatomical textbooks until the last gasp of the twentieth century. In many manuals the clitoris was present in the early 1900s, and subsequently deleted mid-century, suggesting that its omission was perfectly intentional – a subconscious means of denying women sexual pleasure, perhaps. The anatomical bible *Gray's Anatomy* was just one of many to remove the clitoral label from their diagram of female genitalia. Other textbooks significantly undersold the clitoris' size, the extent of its nerve supply, or only deigned to mention the external glans, accompanied by a cursory explanation of its form being merely a 'small version of the penis'.

---

* Falloppio was one of the greatest anatomists of his time and a rather unlikely investigative authority on female reproductive anatomy. He was the first to accurately describe the tubes leading from the ovary to the uterus (the 'trumpets of the uterus', as he called them). They were subsequently named the 'Fallopian tubes' in his honour, although he failed to grasp their function. Falloppio coined the term vagina and also disproved the popular notion that the penis entered the uterus during intercourse. Most ironic of all, for a Catholic man of the cloth, he also developed the world's first prophylactic sheath, as a protection against syphilis. He used a small linen cap drenched in a solution of salt and herbs, and sometimes milk, to cover the glans of the penis. The soggy contraption was held firm by a pink ribbon, so that it would 'appeal to women'.

It wasn't until 1998 that the pioneering Australian urologist Helen O'Connell published the first detailed anatomy of the human clitoris and began a noisy campaign for accuracy in medical texts. The rest of the animal kingdom lags way behind, despite the fascinating wealth of clitoral diversity out there, including an internal bone known as the baubellum (American black bears) and even spines (ring-tailed lemur). But little is known about their tissue structure and function. 'With the exception of humans, rats and mice, we don't really know what the clitoris looks like. But all vertebrates have a clitoris,' Brennan told me.

As well as its form, the position of the clitoris is another point of variation. As we shall discover in chapter eight, female bonobos, our closest ape relatives, have a clitoris that's positioned to facilitate mutual stimulation with other females. In humans the clitoris sits outside the vagina, which seems rather inconvenient when you know that in the majority of mammals the clitoris is positioned inside the vaginal entrance, where it can be easily stimulated by the penis during sex.

This is the case for the bottlenose dolphin. A combination of this vaginal position and the enlarged size of the dolphin clitoris screams sexual pleasure to Brennan. The existence and role of such desire in female animals has long been controversial, but is logical. Sex, like eating, is essential for life. So, why should it not feel good?

A cellular inspection of female genitalia is finally putting the debate around female pleasure to rest, and even revealing how it shapes the evolution of male behaviour and physiology. It turns out even female insects enjoy sex, if it is done right. In the bush cricket, *Metrioptera roeselii*, males possess a pair of curved rods, pointing out of their genital opening. They look a bit like coat hangers and for a long time no one knew their function. It turns out they are for stimulating the female during sex and have thus been named, with all seriousness, titillators. CT scans revealed that the male rhythmically inserts his titillators inside the female during copulation, and uses them to drum against a sensitized internal plate. It's a kind of bush cricket foreplay that allows him to copulate. When a male had his titillators experimentally shortened or the female's senses were chemically blocked she resisted the male's advances.

Brennan is interested in the 'sensory fit' of the female and male parts during sex. Sadly her dolphin cadavers were too degraded to perform the necessary histology. But it's likely that female genital stimulation in mammals acts as a form of copulatory courtship, to induce ovulation or aid in the transport of sperm. In which case, a measure of pleasure might be another way a female subconsciously decides if a male gets to fertilize her eggs or not.

'Sex feels good, right? And some things feel better than others. To me that is an indication of female choice,' Brennan explained.

Danish pig farmers know all about this. They've discovered that artificial insemination is more effective if preceded by manual stimulation of the clitoris, cervix and flanks. So they've taken a practical approach and developed a special five-step sow stimulation routine, with graphic images for guidance. Seduction starts with the farmer stimulating the sow with a fist, moves on to massaging her hips and finishes with the farmer sitting on her back to mimic the weight of the mating male on her. This sow seduction formula results in 6 per cent more babies than going in straight with a cold hard syringe, but may put some off a career in pig farming.

Despite their image as racing Olympians, sperm don't actually have the energetic resources or directional swimming skills to travel under their own steam to the site of conception. They need help. In some primates the degree of sperm uptake has been linked to contractions associated with female orgasm. During climax, the release of the hormone oxytocin causes the uterus and oviducts to contract resulting in the 'upsuck' of sperm, which significantly accelerates their passage to the egg. A study of captive Japanese macaques, *Macaca fuscata*, found that females are more likely to achieve orgasm-like responses when mating with high-ranking, socially dominant males, suggesting preferential sperm uptake for these males.

A recent investigation into orgasm in women agreed that female climax appears to promote conception in humans too. The researchers concluded that female orgasm wasn't a by-product of male orgasmic capacity or a means of reinforcing the pair bond, as suggested by Desmond Morris. Instead evidence suggested that, amongst humans, orgasm is more likely a cryptic means of selecting a high-quality sire for a woman's egg. They propose that, just like those *socially* monogamous fairy wrens and hooded warblers from chapter three, ancestral human females may have pursued a mixed reproductive strategy, choosing partners based on investment potential and then sneaking off during ovulation to have orgasmic sex with high-quality males. A cryptic copulatory mate choice mechanism would allow females to exert some control over paternity, even if their sexual choices were constrained by familial influence or sexual coercion.

AND THE WINNER IS . . . THE EGG

The more we investigate the female reproductive tract, the more ownership a female gains over her fertilization rights, and the more ridiculous the idea of an all-important 'sperm race' becomes.

It turns out that mammalian sperm aren't even capable of fulfilling their biological function without female intervention. They can't actually fuse with the ova without a period of activation known as capacitation. This is under female control and involves chemical alterations of the sperm, probably involving uterine secretions. But guess what? We don't know much more because it's not really been studied. Unfortunately, 'although recognized for more than fifty years, capacitation remains an ill-defined process'.

Capacitation offers yet another opportunity for a female's reproductive tract to influence which sperm 'wins'. But whether a female has biased sperm from males that please her, or impedes the sperm from those that don't, cutting-edge research suggests that in the end, it might be the egg itself that has the final say.

Ova have long been considered the very epitome of female passivity; their large size and sedentary nature, in comparison to the small and mobile sperm, the very source of sexual inequality (as discussed in chapter three). Textbook descriptions of fertilization take the form of a biological fairy tale, with the helpless princess ovum waiting listlessly for her heroic sperm prince to battle his way to her rescue and awaken her from her lifeless slumber.

But there is growing evidence that the egg can actually influence which sperm gets to enter, regardless of which one 'won' the race. Unfertilized eggs are known to release chemo-attractants that essentially leave a trail of biological breadcrumbs to guide sperm in the right direction. Not all sperm react in the same way, meaning there is room for the egg to select the optimal candidate, rejecting those that are genetically incompatible even if they arrived first. The egg makes no concessions for romantic partners, however. In the study of human eggs, in over half of the cases the egg preferred the sperm of a random male and not the woman's partner.

This scientific revelation may not sit well without devoted husbands, but as far as a woman's egg is concerned, all's fair in love and genital war.

CHAPTER SIX

Madonna no more:
selfless mothers and other fictional beasts

*Woman seems to differ from man in mental disposition, chiefly
in her greater tenderness and less selfishness . . . Woman, owing
to her maternal instincts, displays these qualities towards her
infants in an eminent degree; therefore it is likely that she
would often extend them towards her fellow-creatures.*

Charles Darwin, *The Descent of Man,
and Selection in Relation to Sex*

My closest brush with motherhood was an intense twenty-four
hours fostering a wild baby owl monkey in Peru. The experience
left me sleep-deprived, anxiety-riddled and anointed with faeces,
which is all par for the course, I am told.

My maternal adventure took place during a month-long stint at
a remote biological field station on the edge of Manu National
Park, in the depths of the Peruvian Amazon. A full day's journey
upriver from any kind of civilization, this vast roadless wilderness
is home to arguably the richest biodiversity on the planet – much of
it unknown to science – plus a couple of dozen zoology geeks running around like kids in a candy store desperately trying to
document and make sense of it all.

The general policy at the Los Amigos station was to observe and not interfere with nature, which meant that rescuing a distressed animal, even if it was endangered, was considered taboo. But when Emeterio, the Peruvian field assistant, stumbled upon the forlorn cries of a gravely injured baby owl monkey, and then, a few metres away, the grisly remains of its half-eaten parent, the hard-hearted policy was temporarily abandoned.

Of the dozen or so species of primate in this rowdy corner of rainforest, the black-headed owl monkey (*Aotus nigriceps*) wins the prize for being the most enigmatic. These miniature primates, roughly the size of a small squirrel, hide out high up in the canopy and aid their secrecy, as the name suggests, by being the world's only nocturnal monkey. They live in family groups and are, unusually for primates, monogamous. Pairs will have just one baby a year – a ball of dense fluff that fits in the palm of your hand and could have been pumped out of a Japanese kawaii factory – all eyes and devastatingly cute.

Our orphaned baby appeared to have been snatched, and subsequently dropped, by a hawk – the owl monkey's deadly nemesis. It arrived at the camp half dead – dehydrated, limp of limb and with a giant gash down its side already squirming with grossly opportunistic maggots. As we did our best to patch him up, I don't think anyone honestly believed he would make it through the night. He was so tiny and so frail. I remember how the primatology team tried to rehydrate him with an injection of fluids, the medical syringe dwarfing his flaccid body.

Somehow, against the odds, he did survive and so, overnight, the field research team became instant parents to a helpless foreign baby primate. We gave him a name – Muqui after *musmuqui*, the Peruvian Spanish for owl monkey – and wondered what the hell to do next. Muqui's unrelenting cries provided us with some guidance to his needs and he took up residence on our heads.

Muqui was calmest clinging to a mass of hair, which made the prospect of potty-training especially onerous. He slept pretty peacefully atop someone's head during the day, so much so it was easy to forget he was there until you bent over and remembered you had a (now screaming and quite possibly urinating) baby monkey clinging on. Night-time Muqui became an altogether different beast. His behaviour was electrified, and we took turns in babysitting the super-charged night owl. After he'd been at the station for a couple of weeks, night-nurse duty fell to me. This is what I wrote in my diary the following day.

> I have to sleep on my front with Muqui nestled on the back of my neck under my (long) hair. He wakes 4–5 times in the night for milk by clambering on my face and rubbing my ear. He must pee and poop after his feed but it is impossible for him to take on board that he must leave the confines of my bed to do so. He prefers to return to my (increasingly) bird's nest hair, which is far from ideal. Generally he is getting bolder – exploring all night by running over my body and frantically climbing inside the mosquito net. Activity peaks around four a.m. with a frenzy of ear rubbing and hair exploration. This morning I look like Robert Smith of the Cure.

According to Darwin, this should all have been second nature to me. My maternal drive should have kicked in and transformed me into an intuitively wise and selfless nurse. But the truth was I felt quite traumatized by the experience – fretful, out of my depth, exhausted and, for the sake of my defiled and defecated-on hair alone, uninclined to repeat the whole sorry ordeal ever again. I was thirty-nine at the time and wrestling with whether I should be having children myself. My night with Muqui only reinforced my suspicion that I came from a long line of not terribly motherly females. If there was such a thing as maternal instinct, I was pretty sure I didn't have it.

THE MYTH OF MATERNAL INSTINCT

Female animals have long been equated with mothers as if no other role existed. Motherhood is an emotive subject – synonymous with nurture and sacrifice; and, as such, riddled with misconceptions, the most fundamental being that all females are born 'natural' mothers, imbued with an almost mystical maternal instinct that drives them to effortlessly intuit their offspring's every need.

The most obvious problem with this idea is it assumes caring for young is the sole responsibility of the female. In Muqui's case, his mother would have suckled him every few hours. But after each feeding bout she would have driven him away, quite unsentimentally, by biting his feet or tail, leaving his father to take on the role of primary caregiver and the heavy job of carrying him 90 per cent of the time.

The commitment to childcare demonstrated by owl monkey fathers is admittedly not the norm amongst mammals (only one in ten mammalian species exhibit direct male care). This comes down to the fact that female placental mammals are physically responsible for both incubating and feeding their young, which makes it somewhat harder to dodge childcare. Male mammals may have nipples, but with a few notable exceptions – namely two species of fruit bat, some inbred domestic sheep and a handful of WWII prisoner of war survivors* – only female mammals have been

---

* The great evolutionary biologist John Maynard Smith once pondered that 'it is odd that no case of male lactation has evolved'. Could this decidedly eclectic group of milk-producing males be the most highly evolved 'new men' on the planet? Thomas Kunz and Charles Francis first discovered lactating male Dayak fruit bats (*Dyacopterus spadecius*) in 1992 while conducting a bat census in a rainforest in Malaysia. Francis removed a bat head first from the mist net and it looked like a female, with its noticeably enlarged nipples. To his surprise, once he studied the bat's nether regions, it was very obviously a male. In all they captured ten males, which all produced small amounts of milk when squeezed. In the case of Dayak fruit bats, the males have the correct plumbing and physiological capability to lactate, but the amount of milk was only about a tenth of that produced by females. Males were not observed nursing young and their nipples were 'smaller and less cornified' than the females', suggesting they had not been suckled. The phenomenon has subsequently been observed in the masked flying fox (*Pteropus capistratus*) in Papua New Guinea. No one is really sure still why this happens, although it is likely that diet is to blame rather than some sort of evolutionary advantage. Many plants contain phytoestrogens (plant-based oestrogens) which could have stimulated the mammary tissue. It is likely the same story with the inbred domesticated sheep. Diet is also key in the

known to produce milk from them. For many, lactation extends for far longer than gestation and ties them into parental duties for months, if not years (orangutans, for example, have been known to breastfeed for eight or nine years). This responsibility has long been viewed as a constraint on females, a costly energetic handicap that restricts the life strategies available to them. Male mammals on the other hand can cut bait any time after fertilization, giving them the freedom to spend their energy procreating with multiple females and fighting off other males.

Once females are liberated from the physiological responsibilities of pregnancy and lactation, as in the rest of the animal kingdom, dads become much more devoted. Amongst birds, bi-parental care represents the overwhelming majority, with 90 per cent of avian couples sharing the load. Slide back along the evolutionary scale and paternal care becomes not only more common, but customary. Amongst fish, it's single dads that do all the nursing in almost two thirds of species, with mums doing little more than donating eggs before disappearing for ever. Some, like the male seahorse, even give birth.*

It's a similar story with amphibians, which display a range of parental care strategies from single dads to single mums to co-parenting. Take the flashy little poison frogs I'd often see hopping about the rainforest floor in Peru. These tiny toxic amphibians (from the family *Dendrobatidae*) make for surprisingly dedicated parents. Occasionally I would catch one bouncing along the forest

---

case of the WWII prisoners of war; after they were liberated and provided with adequate nutrition, the resulting imbalance of hormones caused them to lactate. Liver cirrhosis can cause a similar condition.

* Following courtship, the seahorse female uses a tubular ovipositor to shoot her eggs into the male's brood pouch, where he promptly inseminates them. New research has shown the male's fleshy pouch is remarkably womb-like: richly vascularized with blood vessels that control the salinity of the developing fry's environment, as well as providing oxygen, nutrition and removing waste gases. This suggests a common toolkit of pregnancy genes utilized by male seahorses and female mammals. After twenty-four days, muscular contractions force the birth of around two thousand baby seahorses. Within hours the male is ready to be impregnated by another female and go through it all again.

floor with a bunch of tadpoles stuck to its back like a wriggling knapsack. This surreal behaviour appears to be quite aberrant, but the frog is actually giving the freshly hatched tadpoles a piggyback to a safe water source. Poison frogs lay their eggs in the leaf litter, but tadpoles are aquatic and so, once they hatch, they need to metamorphose in water, such as the tiny puddles that collect in tree holes or the recessed base of bromeliad leaves. These temporary private pools are predator-free and a safe place for the tadpoles to complete their development without getting eaten. Frogs will travel for hours, if not days, sometimes scaling vertiginous rainforest trees several storeys high, in search of the perfect pool for their kids, which when you are only an inch long is an extraordinary act of parental commitment.

In the wild, this marathon mission is mostly performed by males, but in a few species of poison frog it's the females or even both parents that take on these duties. Lauren O'Connell, assistant professor of biology at Stanford University, recognized that this variability amongst closely related species offered a unique opportunity to examine the neural circuitry controlling parental care, and to find out if it is the same across the sexes.

'I think when people think about frogs they think of them as having totally different brains or wonder if they even have brains at all,' O'Connell told me over Skype. 'Yes, they do! And in fact their brains are quite ancient, so the same pieces are there in all animals. It's just in whether they're bigger or smaller that the complexity differs, but they're still all the same pieces.'

In the dyeing poison dart frog (*Dendrobates tinctorius*) females never piggyback tadpoles in the wild, it is always the male. But in the lab when O'Connell removed males she found the females often (if not always) stepped up and took on the role. When she looked inside the frogs' brains, she found this behaviour associated with the activation of a particular type of neuron in the hypothalamus that expresses a neuropeptide called galanin, regardless of their sex.

'The circuitry facilitating parental behaviour is the same in males and females,' O'Connell told me.

So it's not that one sex is programmed to instinctively provide care, it's just that one does it. But both retain the brain architecture to drive the care instinct. Which is something of a body blow for the idea of a unique *maternal* instinct, in frogs at least. But what about mammals, where the lion's share of parental care is often provided by the female and males can be, well, somewhat less inclined to nurture?

In the mouse world, for example, virgin males are aggressive and infanticidal, regularly injuring or killing newborns. Recent groundbreaking research led by Catherine Dulac, Higgins professor of molecular and cellular biology at Harvard University, has revealed that these murderous male mice can be transformed into doting dads by stimulating the very same galanin neurons in their hypothalamus.

'It's like a switch for parenting,' Dulac told me over Zoom.

Using cutting-edge optogenetic techniques, Dulac was able to activate the galanin cells in virgin males on the verge of killing infants. Their transformation was instantaneous. The males started building nests, and carefully placing pups in them, which they then groomed and huddled protectively.

'The male mice were "maternal" – they took care of pups in exactly the same way a mom would do, the only difference being that they couldn't lactate. It's absolutely fabulous.'

Dulac discovered there are two groups of neurons – one that drives parental behaviour (galanin neurons) and another that drives infanticidal behaviour (urocortin neurons) – and they project directly on to one another. Stimulating one inhibits the other, so the behaviours are mutually exclusive – an animal cannot be both parental and infanticidal.

This neural circuitry is the same in males and females. When Dulac used the same techniques to stimulate the urocortin neurons in female mice they flipped from caring for their pups to attacking them instead. 'It's amazing. You push one button, you have parenting behaviour. Push another button, you have infanticidal behaviour,' she explained.

Dulac had stumbled upon the neural control for the most funda-mental *parental* instinct: don't eat the kids, care for them instead.

It might seem obvious to cognitive species like us that chowing down on your children might not be the most auspicious first move for parenting. But not if you're a frog. If you place random spawn in front of an amphibian they will simply gobble it up – eggs are a tasty, free protein meal, after all. So it makes sense that the neural circuitry for parenting switches off the basic instinct to eat. This impulse is particularly useful for the 'mouth brooders' – frogs and fish that incubate and protect their young in their mouths. Clearly the impulse to care-not-consume would be pretty helpful for this, the parenting equivalent of sucking a gobstopper for weeks on end and trying not to chew.*

According to Dulac infanticide is also common amongst mammals – having been recorded in around 60 per cent of species. Male mice routinely kill babies that are not theirs, but won't if they are the father. Mothers, on the other hand, will sacrifice their own babies if they become heavily stressed – by predators or hunger.

To Dulac, it makes sense that the compulsion to parent must be equally hardwired. Animals are fundamentally concerned with their own survival, and rightly so. Theirs is a life-or-death world, in which the struggle to eat or not be eaten is a daily deadly reality. 'Why would a female mouse take care of a little pink thing that suddenly arrives, screams a lot and is very demanding? Having to handle it and make sacrifices is extremely dangerous,' she says.

---

* This strategy is adopted by a handful of fish (both male and female) and one of my favourite anurans, Darwin's frog (*Rhinoderma darwinii*) – a candidate for Nature's Best Dad. Darwin's frog swallows his clutch of a dozen or so eggs just before they've hatched. They stay in his throat sac for eight whole weeks, after which he barfs up whole baby froglets. For the whole 'pregnancy' he doesn't eat and is muted by the experience. I once travelled to the far-flung dry forests of Patagonia to see one of these devoted fathers in the wild and, after much searching, found him hopping about outside the gents' loo in a remote national park. To my great excitement he was pregnant. His throat sac was alive with tadpoles and looked a lot like John Hurt's stomach just before it erupts in the original *Alien* movie.

'This is where parental instinct is kicking in. It says you have no choice but to care.'

As Dulac sees it, both strategies are essential for the survival of the species. 'You and I are alive today because some of our ancestors used their galanin cells to nurture our ancestors, but also because there were these urocortin cells that allowed mum to decide whether it was a good time to have babies or not. Without them she might have died,' Dulac told me. 'I think it's important to remember that.'

The big question is what triggers this switch to parental care. Dulac has yet to figure that out but her hunch is that it involves a number of profound circuit changes, prompted by a cascade of internal and external stimuli. The galanin neurones then act as a sort of command centre for parenting, coordinating input and output from all over the brain to create a spectrum of caring behaviour that goes way beyond a pre-packaged one-size-fits-all innate response, and results in a huge amount of variation in individual parenting style and ability, irrespective of sex.

'We are so simplistic in the way we see things as being either male-specific or female-specific,' Dulac pointed out to me over Skype. 'If you look around, whether it's in humans or in any animal, not all individuals behave the same. Not all males are equally aggressive. Not all females are equally maternal. There is enormous variability.'

Not only is this neural circuitry the same in male and female mice, but Dulac suspects it is shared by all vertebrates, including us. The hypothalamus is an ancient area of the brain, the centre for a host of hardwired behaviours like sleeping, eating and sex. Each time the neurons controlling these behaviours are discovered in animals, an equivalent has been found in humans. According to Dulac, if this command centre exists in frogs and mice then there's every reason to assume that an analogous circuitry for parenting exists in the brains of men and women.

'What's kind of heart-warming is that when I give talks about paternal and maternal behaviour my male colleagues just love the

idea that the male brain has all it takes to be parental – it's satisfy-
ing somehow,' Dulac said.

A deeper understanding of the full neural circuitry of parenting
may eventually help with the treatment of psychiatric disorders
associated with motherhood. 'What is extremely striking is the testimony of women with postpartum depression; they have these
obsessive thoughts about harming their children. It's extremely disturbing for the women involved. They mostly don't act on them,
but those with psychosis sometimes do,' she told me.

Infanticide is strictly pathological in humans. We are part of the
40 per cent of mammal species that have evolved alternative strategies that eliminate the need for such a blunt survival tool.
Nevertheless the same urocortin neurons still exist in the hypothalamus of humans. They're not used, but Dulac believes they have
been conserved because they've been so important in evolutionary
history. If Dulac is right about the link between these urocortin
neurons and postpartum depression, she hopes that her work can
help identify drugs that could act as blockers to treat such
disorders.

'If you listen to the testimony of women who've killed their children, they don't know what prompted them to do it, but they had
this enormous instinct to do what they did and they have no explanation for it. You can imagine that this might be very similar to
what's happening in the brain of a mouse surrounded by danger
and instinctively deciding to eliminate an infant.'

Dulac's work is the first time a complex social behaviour has
been mapped out in the brain of a mammal. It is a discovery so significant that it won Dulac a prestigious 2021 Breakthrough Prize
(the self-proclaimed 'Oscars of Science'), which goes to show how
academic respect for the study of motherhood has changed in the
last fifty years. The idea that maternal behaviour varies from one
individual to another may seem obvious now, but it wasn't always
the case. When the anthropologist Sarah Blaffer Hrdy went to
Harvard in the 1970s, 'Mothers were viewed as one-dimensional
automatons whose function was to pump out and nurture babies.'

Just as females were thought to be passive and homogeneous when it came to courtship and mating, so it was with motherhood.

'Most adult females in most animal populations are likely to be breeding at, or close to, their capacity to produce or rear young,' wrote the standard textbook of the time, as if to see one mother was to see them all. Natural selection requires variation in order to function, which means mothers were effectively barred from the evolutionary party on account of being too boring.

This ludicrous prejudice was given suitably short shrift by the fiercely analytical mind of Jeanne Altmann, now professor emeritus of animal behaviour at Princeton University and the first scientist to seriously quantify and dignify the evolutionary impact of mothers.

I met Altmann at Hrdy's walnut farm in northern California. The two primatologists are long-term conspirators, although, on the surface, quite different characters. Compared to her larger-than-life Texan host, Altmann, a no-nonsense New Yorker now in her eighties, is diminutive in stature, quiet and reserved. Her ideas have been no less radical, however. Her secret weapon has been an unswerving devotion to the geek god of impartial data. Yes, Altmann sparked a revolution through rigorous statistical analysis, which might not sound terribly sexy, but it was the only way to skew attention away from the dramatic allure of pugnacious male primates.

Altmann could teach logic to Captain Spock. She had started her academic career as a mathematician at UCLA – one of only three women in her class – but was forced to quit when nobody in her faculty felt it worth their time to mentor a woman. Maths' loss was zoology's gain. Altmann arrived at field biology unencumbered by primatology's phallocratic baggage and uniquely qualified to tackle the pitfalls of observational bias, which so plagued the science.

She developed a method of sampling subjects at random and made sure each individual was watched for the same number of minutes – because statistically all behaviour is of equal importance, however 'boring' it may seem to be. The resulting paper outlining her methodology, 'Observational Study of Behaviour: Sampling Methods', was nothing short of revolutionary, for ever changing the face of field research, not just in baboons, but across the animal kingdom. Cited over sixteen thousand times to date, it's been described to me by one anthropology professor as 'inadvertently, one of the greatest feminist papers of all time', since it finally gave females the same airtime as males.

Altmann's second stroke of calculated genius was knowing that she needed to be in for the long haul, collecting data on the same groups of animals for generations in order to calculate the ramifications of their behaviour in deep time. She chose savannah baboons.

Back in the 1960s a young Altmann and her husband Stuart travelled to the foothills of Mount Kilimanjaro in Kenya to study the ecology and social behaviour of savannah baboons (*Papio cynocephalus*) living on the edge of Amboseli National Park. These largely terrestrial and highly intelligent monkeys live in troops of up to one hundred and fifty individuals and had been of interest to primatologists searching for parallels with human society. Altmann's landmark study, which continues to the present day, forged new ground by ignoring the lavish spectacle of competing males that so transfixed mainstream science and focused instead on mother–infant relations. It was a courageous career move. Not only were mothers considered to be of little theoretical significance to the male zoological establishment, they were also out of fashion with female scientists keen to embrace the nascent feminist movement. The study of motherhood was considered a retrograde move: ' "the home economics" of animal behaviour', as Sarah Blaffer Hrdy pronounced in her book *Mother Nature*.

Altmann still travels to Amboseli to check on the baboon project she founded, although she has long since handed the reins over to others. At over fifty years old, it is the longest-running primate study in existence. The accumulation of almost seven generations of unprejudiced observational data on the lives of over 1,800 individual baboons has transformed our understanding of not just the species but mother–infant relations in general. Altmann was the first to provide evidence that primate motherhood is much more than a one-size-fits-all knee-jerk response to nurture, but a multifarious life-or-death business spent negotiating a succession of critical trade-offs, all the while walking a perilous energetic tightrope.

Altmann's research revealed that every baboon mother juggles the demands of a 'dual career'. She must spend 70 per cent of each day 'making a living' – foraging, walking, avoiding predation – whilst at the same time managing the job of caring for her infant. Baboons travel several kilometres every day in search of small fruits and seeds to eat. When a female gives birth, there is no downtime to recover. Though exhausted from the effort, she must keep up with her troop, carrying her infant with one hand and doing her best to maintain pace while walking on her remaining three limbs. If the infant is not carried in the correct position, then it cannot suckle and can quickly dehydrate and die.

Mastering this technique can be especially challenging for first-time mums, which Altmann tells me are often 'puzzled' by their infant's distress. The trigger to parent may be hardwired, but maternal behaviour unfolds gradually – there is much for first-timers to learn. Some are more natural than others. Altmann remembers one young mother whose struggle to nurse had fatal consequences. 'Vee's first infant, Vicki, was not able to get on the nipple during her first day of life; her mother carried her upside down, even dragging her and bumping her on the ground for much of the day.' Although Vee, like most first-timers, picked up the ropes within a few days, it was too late. The damage had been done and Vicki died within a month. Such deaths are not unusual. Amongst primates, mortality rates for first-born infants are up to 60 per cent higher than for subsequent siblings.

Even amongst experienced baboon mothers infant death is a very real threat. Altmann discovered that 30–50 per cent of babies perish within the first year, with nutritional stress being a key culprit. Food is scarce on the dusty plains of Amboseli and the

environment unforgiving and unpredictable. A breastfeeding mother
has to find enough calories for two, which by the time the infant is
6–8 months old is physically impossible to do while carrying her
baby – their appetite being by now too large and body too cumbersome. This creates a conflict of interests between mother and baby.
In order for the mother to survive, her infant must begin to walk and
forage for itself, but infants prefer to continue with the free ride, so
they'll try to manipulate their mothers using 'psychological weapons'. These take the form of truly epic tantrums – the kind that
would put a two-year-old human toddler to shame. Emotional out-
bursts continue until the infant has become fully independent,
somewhere between its first and second birthday.

For mothers, the timing of weaning is a close instinctual calcula-
tion of resources. Mistimed, it can lead to certain death for either
mother, baby or perhaps both. Baboons, like all mothers that breed
several times during their lifetime, must balance the investment in
their current offspring against their own survival and future reproductive capacity. The average baboon is likely to spend 75 per cent
of her life having babies – around seven in total, of which only two
are likely to survive into adulthood. This rather paltry reproductive
result illustrates what a gamble each baby is. Altmann found that
baboon mothers were pushing the envelope of their own survival –
if they were to breed any faster they would risk maternal depletion
and death.

Not all mothers are born equal, however; a female baboon's lot in
life is strictly dictated by her social class. Males may duke it out for
the alpha position in their hierarchy, but females also inhabit their
very own chain of command – a rigid female aristocracy, worthy of
the British nobility. Status is immutable and inherited – passed
down the maternal line – and laced with privilege. Those females
lucky enough to inhabit the upper echelons have first dibs at food
and water sources, they can displace the hoi polloi to get groomed,
and generally enjoy the freedom to go wherever and do pretty much
whatever they want in the group. Including snatching and even kidnapping other females' babies.

Altmann isn't sure why baboons do this. It could be an aberrant side effect of an innate attraction to babies, which is universal in primates. Just the same as it is with humans, a new baby has a magnetic quality. They become a focal point and other baboons want to hold it, especially juvenile females, which are prone to mishandle a fragile baby and quickly lose interest, sometimes with lethal consequences.

'Most primates are all-day suckers,' Altmann told me. 'There's no milk in the kidnapper's breast and very quickly, especially in an arid habitat like Amboseli, the kid will start failing from dehydration and then from lack of nutrition.'

Low-ranking mums are especially vulnerable. They lack the social standing to be assertive so high-ranking female daughters can just grab their baby, and when they get bored of playing with their new toy they abandon it, which can be heartbreaking to watch. 'Some are more clueless than others,' Altmann told me. 'I remember watching one low-ranking first-time mother who would just walk into the middle of trouble. She'd try to keep feeding when others were examining her infant and you wanted to say, "No! No, they're going to kidnap that infant."'

Daughters born into baboon nobility have the advantage of their mother's social connections, which turns out to be the greatest gift of all. This wide network of benevolence offers protection from the threat of kidnapping females or infanticidal males as well as the competitive aggression of other baboons. High-born, well-connected infants are more likely to be tolerated when feeding near by another adult. This support system means mothers don't have to be the be-all and end-all for their kids, which is especially helpful for first-timers surfing a brutal maternal learning curve. Altmann found that daughters surrounded by high-ranking kin give birth at an earlier age to offspring more likely to survive, giving them a lifetime reproductive advantage over mothers in the lower ranks.

This social privilege, or lack thereof, has a massive impact on a baboon's mothering style. Noble-born mothers have what Altmann described as a 'laissez-faire' approach to the job. They have the confidence to let their infants roam far and wide, and exhibit tough love early on when it comes to weaning. This hands-off approach makes for self-sufficient and socially integrated juveniles, which gives them a higher chance of survival as adults.

'The goal of successful parental care is that kids can be independent,' Altmann told me. 'A mother who's not too protective has an infant who explores and develops their social world safely, but independently.'

Low-ranking females are put upon by just about everybody. Without the social standing to protect them and their baby, they compensate with what Altmann describes as 'restrictive' parenting, keeping their infant constantly within arm's length. Restrictive mothering *probably* results in a better initial survival rate, as the infants are safer from predation and disease in the first few weeks. But they develop independence more slowly and place more demand on the mother's critical resources, pushing her closer to that energetic cliff edge, and death.

Faced with a non-stop stress-fest of potential threats, low-ranking mums live life on perpetual high alert. They must watch as their offspring are confronted by dangerous group members, yet are powerless to do anything about it. As their innate warning system screams, their anxiety increases exponentially in the face of social inequality. This stress, detected in the hormones excreted in their faeces, is thought to lower their immune response and make mothers more vulnerable to disease. It can also manifest as depression and even infant abuse. Humans are not the only primates to suffer from post-natal depression. In olive baboons, mothers with low dominance rank were found to display higher levels of abusive behaviour during the postpartum period. In wild populations of macaques, 5–10 per cent of mothers have been observed biting, throwing or crushing their infants on the ground. Some have been known to perish as a result. Those that don't are psychologically scarred and more likely to mistreat their own young, ensuring that this abusive behaviour ricochets down the generations.

Although it might appear that low-ranking baboons are doomed by birth to a life of abject maternal misery – forced to bet their life on the equivalent of a seven-deuce hand in poker – there are ways they can cheat the system and give the next generation a better chance of survival. Altmann's team discovered that if they are able to forge strategic friendships with other baboons, either male or female, they can gain much-needed assistance when running the brutal Darwinian gauntlet.

'We showed that those females who have friends live longer and their kids survive better,' Altmann revealed to me over the phone.

The currency of baboon friendship is grooming, which releases endorphins and helps dial stress levels down on a physical level. As long as a baboon has the time, energy and motivation to start and maintain friendships, they can provide a support network as valuable as family. Less is more, however – it's the strength of the bond that matters, not the number of friends a female baboon has. The ability to forge strong and enduring social bonds may even generate more reproductive benefits than high rank.

'Friends can make a difference in aggression – in eyes and ears for trouble, tolerating feeding close by, finding food sources that they're willing to share. All the things that you'd expect of human friends can happen with baboons,' Altmann explained.

CONTROLLING MOTHERS

Baboon mums have another destiny-cheating tactic: they can unconsciously manipulate the sex of their offspring. Altmann discovered that at her Amboseli study site low-ranking females had more sons than daughters. This plays to their advantage. Female status passes down the maternal line and is fixed, so low-ranking daughters remain shackled to the disadvantage of their mother's miserable status for life. Male baboons, on the other hand, fight one another to dominate their hierarchy, so their status is more fluid. Plus a son who manages to hitch his reproductive star to a more successful matriline and score himself a high-class daughter will secure a ticket to posterity for his offspring. His genetic legacy therefore manages to escape its low-class purgatory. So, if you are a low-status mother it makes sense to produce sons who have a chance to escape the confines of a nepotistic matriline, instead of daughters, who are trapped.

In contrast, high-bred female baboons produce more daughters than sons. Since their privilege is assured, daughters are a less risky gamble and, Altmann observed, have a higher chance of survival than high-born sons.

When Altmann exposed the female baboon's sex-rigging trick, many found it hard to believe such a calculating motherly move was possible. Across the animal kingdom, mothers are way more controlling than you might comfortably imagine – sex-manipulation being just one ploy utilized by mothers from fig wasps to kakapos.* Such unconscious yet biologically calculated favouritism doesn't end at birth. Bird mothers can micromanage the hormonal and nutritional contents of individual eggs to give certain siblings an advantage over others. Mammalian mothers can tailor their milk to

---

* The kakapo's impulse to rig the sex of its offspring to suit the environmental conditions it's born into almost thwarted the effort to conserve them. These strange flightless parrots from New Zealand have been of concern for many years: in 1995 there were just fifty-one of them. Scientists were so worried that they gathered them up and put them all on nearby islands with no mammal predators (invasive rats and feral cats being their greatest threat). However, even in this safe space and with regular feeding by the scientists their numbers had still not flourished and in 2001 still only eighty-six remained. What was going on?

   It turns out the kakapo population had become male-biased (not good for population growth, which is female-driven). If resources are low, mothers will produce females; they are smaller, take fewer resources and have a consistent chance of mating. However, if resources are abundant, mothers will switch to producing males as, even though they are more expensive to make, large healthy males outcompete other males and produce more grandchildren. As the scientists were providing the population with a large amount of food, most females were producing males. José Tella discovered this in 2001 and now conservation efforts have calculated the right amount of food to provide the population to ensure a 50:50 sex ratio.

fit their offspring's specific requirements. Macaque males, for example, get a shorter burst of richer, high-density milk, whereas females get weaker milk but are nursed for longer. Rich milk may help sons bulk up faster and give them the much-needed competitive edge as adults.

Exactly how baboon mothers, be they low- or high-ranking, manage to fix the genetic deck in favour of the 'right sex' isn't yet clear. But in other sex-manipulating mammals, like coypu and red deer, their method is strategic abortion.

The idea of a female taking such brutal command of her reproductive destiny may not be welcomed by the pro-life brigade. The uncomfortable truth, however, is that nature is decidedly pro-choice. Abortion, at every stage of pregnancy, is an unconscious adaptive strategy for many animal mothers facing unfavourable situations that place themselves, or their offspring, in peril. Even pandas do it. I spent a summer waiting to film the much-lauded Tian Tian giving birth at Edinburgh Zoo, only to be informed at the final hour by the resident press officer that, unaware of the world's watchful eyes or the potential TV ratings, she'd had the temerity to 'reabsorb her fetus'. This is a common, and decidedly prudent, ursine solution for dodging a doomed motherhood in stressful conditions (and as an unexpected bonus also saves her cub from a life sentence behind bars).

In the wild, gelada baboons will abort when a new dominant male takes over the group. Incoming males almost always kill any babies they didn't sire, so terminating the pregnancy is a mother's insurance policy against this almost inevitable infanticide and wasting any further reproductive effort on a likely dead end. Known as the 'Bruce effect', after Hilda Bruce, who first discovered the phenomenon in mice around half a century ago, this particular trigger for abortion has since been documented in all sorts of wild mammals from lions to langurs.

The goal of motherhood isn't to nurture babies indiscriminately but for a female to invest her limited energy in creating the maximum number of offspring that survive long enough to reproduce themselves. There is nothing truly selfless about the job; it is absolutely selfish. A 'good mother' instinctively knows when to sacrifice all for her offspring and when to cut bait, which might even be after the infant is born.

In the badlands of the Australian outback, the female kangaroo has evolved an ingenious way of hedging her bets in the face of a capricious environment – a reproductive assembly line that allows her to simultaneously juggle progeny at three different stages: a suckling but almost independent joey that spends little time in the pouch and hops by her side; a pink jellybean of a joey fixed to a nipple in her pouch; and a fertilized but dormant ball of cells known as a blastocyst frozen in suspended animation in her uterus. When pursued by a predator, she can jettison the bigger joey from her pouch, lightening her load and allowing her to escape. Unable to keep up with its mother, a teenage joey will perish without milk or protection from its mum. While this may sound heart-breaking to a human, for the kangaroo no painful conscious decision-making is required: natural selection has already furnished her with a functioning plan B. The cessation of suckling will trigger her embryo-in-waiting to emerge from developmental dormancy and serve as a rapid replacement for the lost joey.

Far from being evolutionarily insignificant, motherhood turns out to be a high-stakes gambler's game with the potential for great wins or mortal losses depending on how skilfully the venture is played. From an evolutionary standpoint, male dominance comes and goes in a fist fight. Sure, it creates big winners and losers in terms of how many females a single male can inseminate. But the influence of a mammalian mother can be felt for generations and goes way beyond contributing just 50 per cent of the genes. Altmann and her team demonstrated how a mother's social standing affects the actual expression of her infant's genes along with their own social development. This is hugely impactful. Motherhood may be energetically demanding, but viewed another way, that high price tag offers female mammals far more control than males in shaping their precious genetic investment. Viewed from this alternative perspective, far from being irrelevant, mothers actually make a greater evolutionary impact than fathers. And the way Altmann sees it, that gives them greater power too.

'With mammals, the female is stuck with this infant and the infant is stuck with the mother and this was traditionally seen to be a major constraint,' Altmann explained. 'People focused on the constraint, but that's just a piece of this story. It also provides an asymmetry in power in who influences the Next Generation, and I think that's still gotten much less attention.'

Generation after generation of primate mothers are quietly competing for more enduring stakes than the isolated copulations the males are noisily fighting over. The tentacles of maternal control may even extend as far as manipulating the outcome of these highly documented male sexual contests and conquests, in previously unforeseen ways. Recent research has shown that high-ranking bonobo mothers act as matchmakers for their sons, brokering their sexual career with their own status and making it three times more likely they become fathers.

The work of Altmann and Hrdy paved the way for such revelations. They promoted primate mothers from 'plodding constants' – whose steady reproductive output was something of a dreary inevitability – to become equal players, alongside males, in the evolutionary game. Their vision of 'the good mother' challenged that of the natural Madonna and replaced her with a more authentic and complex female figure that's ambitious, calculating, self-seeking and sexually assertive.

The intense drive to nurture and protect remains a key part of that maternal mix. There's no denying the transformative power of motherhood to conjure a deep and profound connection between two naturally selfish strangers. The mystical mother–infant bond is very real indeed, if not as omnipresent or instantaneous as Darwin would have had us believe. I travelled to an uninhabited rocky islet off the east coast of Scotland to discover the powerful yet precarious hormonal scaffolding that underpins this iconic relationship.

FROM FEAR TO MATERNITY

Dawn on the Isle of May and I felt like I was in a zombie movie. The emerging sun had painted the sky a rich blood red but had yet to illuminate my chilly surroundings. Nevertheless, I knew I was not alone. The biting wind was haunted by a cacophony of baleful wails, ominous gurgles and snotty snorts. In the gloomy dawn light I could make out large lumbering shadows – over two metres long – all around me. I'd been warned to keep my distance. The hefty brutes were hostile, armed and highly aggressive. If I strayed too close their first warning would be to hurl a dollop of fishy phlegm at me (evidence of which could be seen glinting on the rocky ground beneath my feet, its slippery nature providing its own additional hazard). Their second line of defence would be considerably more deadly: a bite that could make my arm drop off.

As the sun took hold of the sky the monsters were revealed: hundreds of velvety grey seal mums with soulful dark eyes and their insanely cute snow-white fluffball pups. Every November the Isle of May is transformed for three bellicose weeks into a maternity ward for around four thousand grey seals (*Halichoerus grypus*, meaning 'hook-nosed sea pig' after their patent Roman profile). It's a riot of aggressive maternal love.

Grey seals spend most of the year as solitary aquatic hunters, but once a year these antisocial beasts must haul themselves out of the water to give birth and nurse a single pup in the company of strangers. This scrap of storm-battered rock is the largest east coast breeding colony of grey seals in Scotland and, at just 1.5 km long and 0.5 km wide, a congested and pugnacious pupping party.

'People think grey seals are cute,' Kelly Robinson told me, 'but as a researcher you learn not to go near the bitey end.'

When I met Dr Robinson in 2017 she was a young researcher at the University of St Andrews, one of twenty or so zoologists drawn to the Isle of May as part of the Sea Mammal Research Unit. This long-term study group have been chronicling the rowdy spectacle of seal pup season for decades, since it offers a unique opportunity to study maternal behaviour in a large mammal at remarkably close quarters.

Such work is not without risk. Seals are surprisingly fast for an animal that's forsaken legs for flappy little flippers. They have teeth worthy of their apex predator status, powerful jaws and a mouth teeming with harmful bacteria. Over dinner on my first night at the island field station, Robinson and the rest of the team regaled me with terrifying tales of a virulent and potentially lethal infection caused by contact with seal bodily fluids, with the ominous name of 'seal finger'.

'The story goes, you should cut your hand off rather than risk getting "seal finger",' Robinson warned me. Legend has it, the infection from a seal bite or contaminated wound can spread so quickly up your veins that amputation becomes the only sensible option. Once the scourge of seal hunters, it's now the seal research-er's greatest bogeyman. Although tourists should also be wary. Robinson told me of a recent medical paper documenting a unique case of 'seal buttock' after an elderly man on an Antarctic cruise got chomped on his right cheek while retreating from an angry fur seal on Deception Island. 'Seals are big and angry and not play-things for tourists,' Robinson concluded.

Robinson has had to become highly adept at avoiding 'the bitey end' of grey seals, since her work involves taking occasional blood samples as part of her investigation into the hormones behind the mother–pup bond. The onset of motherhood is facilitated by a symphony of hormones, but there's one that stands out as a power-ful driver of the maternal experience: oxytocin.

You may well have heard of oxytocin. This famously feel-good neuropeptide has gained considerable notoriety as the 'cuddle' or 'love' hormone on account of its magnetic power to foster attachment. It works in conjunction with the mood-altering dopa-mine reward system to deliver highly addictive warm, fuzzy feelings in a wide range of relationships, not just mother and child. The afterglow of sex – that's oxytocin encouraging you to bond with your sexual partner, however inappropriate. Oxytocin is the glue that keeps the unusually monogamous prairie vole faithful to its partner for life. When chimpanzees groom one another, they release oxytocin, strengthening friendships and alliances. It is even released when I gaze at my pet dog. Oxytocin is the MDMA of the hormone world, but don't be fooled by the cosy branding – this complex neuropeptide is much more than just an elixir of contentment.

'Oxytocin is fundamentally involved in the actual physiological process of becoming a mother,' Robinson explained to me. It acts as a smooth muscle contractor, which, in mammals, stimulates the uterus to pump out babies – hence its name from the Greek meaning 'swift birth' – as well as the nipples to eject milk. The physical process of labour is stimulated by oxytocin in the bloodstream, but the stretching of the cervix and vagina during the birthing process itself triggers an almighty rush of oxytocin in the brain. The resulting delicious cocktail of natural opiates ensures the new mother is primed to bond with her newborn as soon as it enters the world. The act of suckling will bathe her brain in yet more oxytocin, so she basically becomes addicted to caring for her baby.

'Oxytocin influences behaviour, but it's also influencing all these other physiological actions and without both halves of that equation you wouldn't be a mother,' Robinson explained. 'So there's that key linkage, which for me is really interesting because it's highlighting how nothing evolves in isolation.'

This flood of oxytocin rewires a mother's brain, tuning it to the cries, smells and sight of her offspring. It seems to make social information more salient by connecting brain areas involved in processing social information – whether it's faces, sounds or smells – and those areas to the dopamine reward system. So when her baby cries for the hundredth time, a mother will be motivated to respond by the delivery of yet more natural opiates.

'We think the mother's oxytocin is one of the few positively reinforced hormones. The more oxytocin a mother releases from, say, milk let-down, the more it triggers oxytocin release,' Robinson told me.

Lactating mothers are certainly drunk with the stuff, which could explain their heroic levels of self-sacrifice. Mothers can be famously fearless when it comes to defending young – 'never come between a mother bear and her cubs', as the old adage goes. This heightened ferocity is unique to nursing mothers, and the so-called 'cuddle hormone' has been implicated in their Jekyll and Hyde transformation. Oxytocin is thought to reduce anxiety and fear, helping a mother cope with maternal stress and priming her to val-iantly defend her offspring against any potential threat.

A strong maternal bond is essential for grey seals as their pups are totally dependent on their mother for nourishment. True seals have the shortest lactation and fattiest milk of any mammal. Grey seal pups are weaned in just eighteen days on milk that is 60 per cent fat. During this time the mother cannot return to the sea to feed so she loses up to 40 per cent of her body weight. Her pups, meanwhile, treble in size.

'They go from being these super-tiny thin pups to being this mas-sive round ball of fat,' Robinson told me. 'I've seen weaners roll down hills and they can't stop themselves because their flippers can't touch the ground.'

The 'weaners' wait up to a month before they leave the island to face life at sea. And no wonder. This is the most dangerous phase of their life, when most will die. The young seals have to learn how to hunt and forage for themselves now. This is no easy task and requires much trial and error. The fattest pups have the greatest chance of survival so it is critical their mothers have stayed close and nursed them well during their short lactation period.

Not all pups become rolling balls of fat. During my visit I saw at least one dead pup – a forlorn, deflated white sock – most likely the result of abandonment by its mother. The Isle of May has more than its fair share of careless mums. Starvation, as a consequence of long-term separations, accounts for 50 per cent of pup mortality on the island.

'I've seen mothers that give birth to their pup and literally aban-don it the second it comes out of them. Their pup is trying to interact with them and they'll just ignore it and roll over,' Robinson told me. 'Everything that the grey seal mother has to do for her pup is crammed into this brief eighteen-day period so there should be high selection pressure for only the best maternal care. So what is going on?'

Robinson found that oxytocin levels provided a reliable metric for predicting maternal behaviour in wild grey seals: mothers with high levels spend more time snuggling up to their pups and developing a strong attachment. Low oxytocin levels suggest a less secure bond, and the female in Robinson's study with the lowest levels did indeed abandon her pup on the fourth day. Her oxytocin level was so low, in fact, it was comparable to that of a non-breeding female. In other years the same mother went on to successfully wean her pups to full term, so what happened on this year to change her pattern of maternal success?

It seems that for mammals that recognize their young by smell there is a critical window, a few hours immediately after birth, when the brain's olfactory bulb has a heightened sensitivity and the mother–pup bond is cemented. If the mother gets distracted during this critical period, say by chasing off a seagull that's scavenging on her freshly birthed placenta, then oxytocin levels can remain at the level of a non-breeding female.

'If for any reason they miss that bonding window, you can't recreate it outside of giving birth or having a massive artificial dose of oxytocin injected into your brain. And that's when you start to see these rejections,' Robinson told me.

Robinson's study is the first to look at the oxytocin levels in pups as well as their mothers and find evidence of double feedback loops in both bonded individuals. Just as the mother's levels are positively reinforced by caring behaviour, her pup's oxytocin levels are similarly boosted by the receiving of care. The close relationship bolsters each other's levels, and the result is that seal mothers with high oxytocin levels have pups with high oxytocin levels. This has dramatic implications for the health and survival of the pups. Robinson discovered that these high oxytocin mother–pup partnerships resulted in the fattest pups, but without costing the mother additional calories (so it can't just be that they're drinking more milk).

'If a pup's got high oxytocin levels then it should be motivated to seek out its mother and stay with her, which means spending less energy running around the colony getting into trouble. The pup could also be getting some sort of microclimate benefit from sheltering right beside its mother in these cold colonies.'

As well as directing pup behaviour, Robinson thinks oxytocin could well be influencing how the pup's fat tissue actually develops, as well as being involved in regulating its appetite and energetic balance. Whatever its methods for promoting weight gain, it's clear that high oxytocin pups have a greater chance of survival.

They may even go on to make better mothers. In the same way that oxytocin helps reconfigure a mother's brain to be more receptive to babies, research suggests it also shapes the gene expression and neural development of offspring. There is evidence in rats that the oxytocin levels experienced by nursing offspring affect their maternal style as adults – those with attentive mothers will go on to make attentive mothers. Variation in nursing style can also affect other social bonds in later life. A lack of infant care in prairie vole pups impacts the density and expression of oxytocin receptors in their brain, which results in compromised social behaviour as adults. Prairie voles are normally monogamous but those neglected as pups failed to forge lifelong sexual bonds as adults, as well as displaying disrupted parenting skills, likely due to increased anxiety.

Robinson's work highlights the long-lasting survival and fitness implications of a strong mother–pup bond. But also its precarious nature. Something Robinson was all too aware of when she recently went through the whole process for the first time herself.

'My baby arrived early – I had to be induced – so I was actually given oxytocin myself, which caused no end of hilarity for my husband and all my friends, who were delighted that I was now getting my own oxytocin manipulations,' she told me.

Robinson is confident that the oxytocin double feedback loop she discovered in grey seal mothers and pups should also exist in humans. There is evidence that, following birth, human mothers demonstrate a unique ability to recognize different sensory cues – visual, sound and smell – from their own babies. In one experiment, mothers with an insecure attachment to their baby were shown to have low oxytocin levels. When shown photos of their babies crying, their dopamine reward systems didn't fire in the same way as they did in women with secure baby bonds. Instead their brains showed increased activation in the area associated with unfairness, pain and disgust.

'Humans are aware that we need to look after our babies. But hormonally, if you're not in the right place to have those behaviours come out, then things can get very difficult,' Robinson admitted. 'My baby struggled to put on weight and through that pressure, that upset and that strain, understanding the underlying process was really helpful to me. There are a lot of misconceptions about motherhood. There's this assumption that there's one optimal road and if you're not doing that, then it's not right. The fact is, life is messy and those ideal scenarios that enable you to do that optimal behaviour are just not going to happen all the time.'

Oxytocin has gained a lot of glory in recent years. But Robinson is keen not to overplay its role as the be-all and end-all of social attachment. It is dangerous to attribute such almighty power to a single molecule, especially in complex cognitive creatures like humans. Fortunately, the biology of maternal bonding doesn't only rely on a precarious rush of oxytocin following birth and lactation. Evolution has ensured there are other longer and more secure roads to attachment, which make caring for babies a more egalitarian arrangement.

PARENTAL CARE – A COMMUNAL AFFAIR

Catherine Dulac is investigating the impact of oxytocin on the galanin neuronal hub – the switch for parental care in both sexes that we met earlier in the chapter. She's found that this parenting command centre does indeed have oxytocin receptors, but only in mothers. This accounts for a birth mother's unique souped-up parenting response – she has both galanin and oxytocin neurons driving her behaviour. But the cuddle hormone, despite its reputation, is not the elusive trigger for this parenting switch, it merely complements it.

Dulac believes there is also a second, long-term phase of attachment that's independent of the hormonal flood associated with birth and lactation and may not be driven by oxytocin alone. This second phase can drive the attachment in mums, dads, other more distantly related kin and even foster parents. It has been observed experimentally in virgin female rats, which are normally extremely hostile towards pups, ignoring or even devouring any they happen upon. But if a virgin is repeatedly exposed to pups, especially if she has a mother to learn from, this inexperienced au pair from hell stops killing and starts nurturing until eventually she's as attentive as a birth mother.

'Foster parents can be as good as biological parents, without having given birth. This is probably due to oxytocin, as well as other neuropeptides,' Catherine Dulac told me.

Humans and rats are not the only species to foster unrelated young. Adoption has been recorded in at least 120 mammals from elephants to shrews. The abandoned grey seal pup from Robinson's study was, in fact, rescued by another female in the colony: an experienced mum already primed for maternal care, which ensured the pup's survival.

Motherhood is an intensely demanding job, with great evolutionary impact. This flexibility of attachment takes the onus off mum to be the only parental figure and allows for a much wider circle of care. Sometimes this is accidental, as in the case of the grey seal adoption, but in other species parenting has evolved into a communal affair, with huge benefits for 'dual career' animal mums.

Bats, for example, can't fly and forage with a baby on board. Instead they make use of a designated nursery and even suckle each other's pups. Giraffes also creche their babies. Foraging adults are somewhat conspicuous and, with their heads in the trees, unable to keep watch over their young while they eat. So they deposit their babies in a nursery, some distance from the main group, with one designated guard. If danger approaches, in the shape of a lion or hyena, the sentry can escort the gaggle of baby giraffes to safety. Hunting mothers also practise collective care. Amongst canids, such as wolves or wild dogs, usually only the alpha male and female in a pack reproduce, but younger group members hunt with the mother and return to the den to regurgitate pre-digested meat into the mouths of her pups.

Although caring for and provisioning someone else's offspring seems at first to defy evolutionary logic, cooperative breeding has evolved many times in a taxonomically diverse array of species. Around 9 per cent of the ten thousand living species of birds and 3 per cent of mammal mums get much-needed help from what are referred to as allomothers, literally 'other mothers'.

I travelled to a far-flung corner of Madagascar to meet a primate mother that has infant day care nailed: the black-and-white ruffed lemur, *Varecia variegate*. These ancestral primates are unusual in that they give birth to litters of up to three babies. Most primate mothers – be they monkeys or apes – generally manage just one at a time. That's because most primate babies, with their big brains, take a long time to reach independence and demand years of intensive care. This makes having more than one at a time too tricky. *Varecia* females have an innovative solution to the problem: they make like birds and build nests high up in the canopy. These serve as communal creches for two or three separate litters, so working lemur mums can share the parenting load.

The biological anthropologist Dr Andrea Baden has been investigating this practical solution to primate childcare for the last fifteen years. When she invited me to join her at her study site in Ranomafana National Park I jumped at the chance, not realizing quite how much effort was involved in decoding the secrets of lemur kindergarten.

Black-and-white ruffed lemurs are critically endangered and can only be found in the few remaining patches of primary rainforest which have yet to be raped of their tallest trees for timber. These fragments of pristine jungle are now only found in the few places loggers have yet to infiltrate, thanks to their distance from roads and challenging topography. I had to hike for twenty-six kilometres, weaving my way through endless rice paddies, baking under a blistering African sun, and then scale what felt like interminable steep, slippery tracks deep into the gloom of the montane forest. It took a whole day's trudge from dawn to dusk, and by the time I arrived at the makeshift camp I was semi-delirious and extremely grateful for my dinner of rice and rubbery, if almost rancid, zebu jerky (desiccation being the only way to store much-needed protein in the absence of electricity).

It was the start of Baden's study season and the lemurs had yet to breed. The young American assistant professor's focus was on tagging as many individuals as possible so she could follow their parental journey high up in the canopy. This involved locating and then chasing lemurs for hours as they cavorted a hundred feet up above us, before darting one with anaesthetic and catching it. All in all quite a workout, especially for the Madagascan lad who clambered to the top of the tree to retrieve our darted and now dormant quarry.

My reward was an unexpectedly close-up encounter; I got to hold my distant primate cousin in my arms and carry her back to camp. Her warm body was about the size of a large domestic cat. Her thick monochrome fur felt exquisitely soft to touch and she smelt of maple syrup – a surprising sugary-scented side effect of her fruity diet.

Tagging individuals is the only way that Baden can follow the lemur's lofty antics. Males and females are identical and discerning individuals at a distance is utterly impossible. Through months of painstaking radio-tracking over several years Baden has nevertheless discovered how 'super-weird' these primates really are. They don't live in fixed groups, like Jeanne Altmann's baboons, or indeed most primates. Instead around twenty-five to thirty adults will have a loose association over a shared territory. 'You don't ever find everybody from the community in the same place together at the same time. They're sort of like atoms bouncing off each other,' Baden explained.

Despite this fluid social life females give birth all at the same time, likely triggered by a bountiful harvest. Such fruity abundance doesn't occur every year, and the lemurs may go for six years without having babies. When they do, they make up for it by having litters. The babies are unusually helpless for primates – blind and unable to even cling to their mother. The first month or so they stay in a natal nest with mum exclusively. Then, once they are big enough, she parks her babies in a communal nest that's close to a big fruiting tree.

'You'll have two or three litters together in a shared nest and sometimes one mom will stay behind with the babies, and the other mom takes off. But what's actually more common is that both moms take off and somebody else stays behind and watches the kids,' Baden told me.

These sentinels keep their babies safe from harm in their high-rise nursery; playful primates are prone to falling out and need to be watched. As well as rescuing tumblers, babysitters also play with their charges, groom them and possibly even suckle them too. Occasionally this guard duty falls to aunts or sisters, but Baden has found that friends – both male and female – are equally, if not more, important. Trust is key and she's recently discovered that females will travel long distances in order to nest with reliable mates. This is borne out by how the mothers spend their time while babies are parked in the nursery. Baden was surprised to discover that while some of their free time is spent gorging on nearby fruit trees, lemur mums are also spending much of their time socializing with other females.

'The saying "it takes a village to raise a child" is really true with these primates,' Baden said, drawing parallels with humans. 'It's evolutionarily significant to rely on others to share the burden of infant care. My mom was a single mom so I really appreciate that. I think community is really valuable and working together with child-rearing is important.'

Sarah Blaffer Hrdy is convinced that this sort of communal care played a key role in the extraordinary evolution of our own species. Human babies really take the biscuit when it comes to being slow and costly to mature. A study of South American foragers estimated that it takes around 10–13 million calories to rear a human from birth to nutritional independence – far more than any single female, however good at spotting big juicy tubers, could feasibly provide.

Whereas Darwin proposed that provisioning by increasingly skilful male hunters subsidized our leisurely development, leading to the evolution of 'greater intellectual vigour and power of invention of man', Hrdy believes that, much like the black-and-white ruffed lemurs, a less gender-specific gaggle of helpers was around to share the caring load. And this maternal help was the real key to our extraordinary intellectual advancement.

In her 2009 book, *Mothers and Others: The Evolutionary Origins of Mutual Understanding*, Hrdy cites a catalogue of evidence from surviving traditional cultures that suggests our ancestors in the Pleistocene may have had a significant degree of help – from men who thought they just might have been the father, to actual fathers, post-menopausal grandmothers, non-breeding aunts and older children. The help from these allomothers is the reason our species was able to foster big brains yet still proliferate.

'Human infants are born larger and even more helpless than any of the other apes, and yet when you compare the birth intervals in modern human hunter-gatherers with the birth intervals in any of the other great apes, you find that babies are weaned sooner and their mothers are breeding at a much faster pace than great apes,' Hrdy has said.

An orangutan mother gets no help and as a result can only afford to have a baby every seven to eight years. By contrast the interbirth interval for a human hunter-gatherer is just two to three years.

Hrdy goes on to argue that this shared guardianship favoured off-spring that were good at soliciting care, thereby promoting the evolution of our unique capacity for empathy, cooperation and understanding the minds of others. In Hrdy's version of human evolution it is sharing the caring load, not hunting and warfare, that shaped the collaborative might and brainpower of the emotionally modern human.

Darwin's maternal instinct lies dormant in all of us. It is not exclusive to females, nor as instantaneous and knowing as the great man would have had us believe. It takes time to awaken and advances in baby steps as we learn the ropes. But it gives us all the opportunity to care for our fellow creatures, with 'greater tenderness and less selfishness'.

CHAPTER SEVEN

Bitch eat bitch: when females fight

Late afternoon on the grassy plains of the Maasai Mara. As the orange sun slinks towards the horizon a pair of topi (*Damaliscus lunatus jimela*) are duking it out in the long shadow of an acacia tree. It's rutting season and the two mid-sized antelope – think souped-up goats on stilts – have joined hundreds of other topi to spar for sex.

The horny pair face off, charge and drop to their front knees as they lock their lyrate antlers, heads wedged against the ground in a vicious stalemate. After a few tense seconds, the more dominant topi maximizes a slight size advantage and shoves the other backwards along the ground. With the ignominy of a sumo wrestler expelled from the ring the loser scuttles back into the herd, shaking its head, leaving the victor free to claim the prize – sex with the prime bull. These armed and aggressive rivals aren't males fighting over cows. They're females duelling over the top topi's sperm.

When Darwin outlined 'the law of battle' sexually combative female topi did not feature. In his reading of the animal kingdom, females have no need to fight over sex. Darwin's theory of sexual selection is, with the rarest of exceptions, all about males clashing over conjugal rights to females. 'It is certain that amongst almost all animals there is a struggle between the males for the possession of the female,' he tells us. With trademark diligence, he goes on to devote dozens of pages to in-depth eye-witness reports of all sorts of macho brawling, from 'timid animals' like moles to 'jealous' sperm whales, all engaging in 'desperate conflicts during the season of love'.

This battle for females explained the evolution of elaborate traits that were otherwise superfluous to the everyday struggle for survival. Costly armaments like horns or ornaments like the peacock's tail evolved expressly for romantic competition and the acquisition of sex. The existence of these so-called 'secondary sexual characters' on the 'passive' female was therefore something of a mystery to Darwin. Horns are just as costly for the female to grow as they are for the male, after all. So why do the females of some species have them?

The idea that females might be using their horns to fight other females is never considered in the many pages of Darwin's thoughtful conjecture. Instead he concludes that although such armaments must be a 'waste of vital power', their presence or absence in females depends not on them being 'of any special use, but simply in inheritance'. A female's horns are thus the ghosts of male bravado, squatting on her head in a redundant fashion until natural selection gets around to removing them.

According to Jakob Bro-Jørgensen, an evolutionary ecologist at the University of Liverpool, such old prejudices die hard. 'When biologists talk about the "battle of the sexes" they often tacitly assume that the battle is between persistent males who always want to mate and females who don't,' he has said.

Bro-Jørgensen is the world's leading (if not only) expert on the sexual politics of the topi. He has observed their annual rut for the last decade and uncovered a complex culture of bitch fights, con artists and bashful bulls that Darwin never dreamed of.

After the short rains in March, female topis travel in large groups to cruise for mates at a lek – the same kind of mating arena we saw the sage grouse using. In the case of the topi, up to one hundred males gather to mark out small adjoining territories. This they do using their own dung, a surprisingly versatile material which creates a signature scent boundary between rivals.

Breeding season is intense, as the females all come into oestrus for just one day of the year. This short fertility window leads to a twenty-four-hour frenzy of sexual activity. Bro-Jørgensen calculated that each female mates, on average, with four males, while some reached as many as twelve different partners in this limited timeframe.

As the females shop for sex, rejected males are not above resorting to underhand tactics to win their attention. If a female leaves his territory unmated, a scorned male will often sound an alarm call, a loud snort that signals a hyena or lion is near by. This fake news alert encourages the exiting female to linger longer in the pretender's territory for safety. With limited time to spare, she often winds up being mounted by the fraudster while she waits.

Bro-Jørgensen calculated that on 10 per cent of occasions topis only succeed in mating by bogus snorting. While some bulls must lie to get laid, others are fighting off females and all shagged out. The top bulls command terrain at the centre of the lek, and it is these studs that have females battling over their limited reserves. 'It is not uncommon to see males collapsing with exhaustion as the demands of the females get too much for them,' Bro-Jørgensen said.

These top males don't just run out of energy, they run out of sperm too. As we discovered in chapter three, contrary to the wishful beliefs of veteran evolutionary biologists, sperm supplies are far from cheap and limitless. Bro-Jørgensen found that females fought violently over the finite seminal supplies from the most sought-after members of the opposite sex. Some pushy cows even went so far as to charge top studs in the act of mounting other females. This brazen tactic doesn't always pay off. Disrupted males will often counter-attack belligerent females and rebuff their advances with added aggression, especially if they've mated with them already.

Bro-Jørgensen discovered that male topi hotshots don't mate indiscriminately, as Darwin would have predicted, but have instead adopted the female's traditional choosy role in order to conserve their precious sperm. Their goal is still to mate with as many individuals as possible, but they deliberately choose the females they have mated with the least to maximize their chances in sperm competition.

It's Bro-Jørgensen's instinct that his topi aren't the only female mammals vying over limited sperm supplies. Such role reversals – antagonistic females and/or picky males – might prove to be widespread, especially among promiscuous species where multiple females favour a few males. 'We may not have our eyes open to the fact that opposite sexual conflicts may occur more commonly than we think,' Bro-Jørgensen said.

Recent research on western lowland gorillas (*Gorilla gorilla gorilla*) supports this hunch. Studies on both wild and captive populations found the females in a silverback's harem compete with one another using sex itself as a weapon in the sperm wars. As we've discovered, silverbacks have notoriously small testes for their body size, suggesting limited supplies. In captivity high-ranking females have been observed having sex outside of oestrus, when there is no chance of becoming pregnant, in order to raid the silverback's semen stores so he only shoots blanks into their low-ranking love rivals. In a wild population in the Congo, high-ranking female gorillas were seen to be even more audacious: harassing, interrupting and even replacing lower-ranking females during copulations. The researchers concluded that tactical non-conceptive sex was an effective 'spiteful strategy' to monopolize the silverback's sperm and resources for their offspring alone.

The revelation that female antelope and ape are scrapping over sex like a Saturday night on *Geordie Shore* has only come to light in the last decade or so. Darwin acknowledged the existence of a 'few anomalous cases' of competitive females in 'reversed' sex roles which 'properly belong to the males' but these were cast aside as trivial exceptions.

Darwin's blinkered yet highly influential view ensured that for the next hundred and fifty years studies of intrasexual competition focused on male competition for mates, and the combative potential of females was largely ignored by science. The resulting data gap on females then masqueraded as knowledge. It's assumed females aren't competitive, and theories are based upon that understanding – when the truth is we just haven't been paying attention.

Birdsong is a case in point. The melodious calls of songbirds have long been considered a classic example of sexual selection: a male ornament that evolved to be ever more elaborate in order to successfully compete with rivals to win the affections of the opposite sex. Birdsong might not seem costly, but memorizing all those songs requires a bigger brain, which is both energetically and physically expensive for a small creature that flies for a living. In fact, the male songbird's brain is known to shrink in the winter months when he's not required to sing.

'Female birds, by selecting, during thousands of generations, the most melodious or beautiful males, according to their standard of beauty, might produce a marked effect,' wrote Darwin in *On the Origin of Species*.

Much like Victorian ladies at a dinner party, female birds had no reason to compete. Hushed by Darwin's theory, their primary role was simply to listen to the jazzy showmanship of the cocks and reward their chosen favourite, albeit reluctantly, with sex. Any female songbirds caught singing were written off as babbling freaks. Their calls fell on deaf scientific ears and were brushed aside with all too familiar excuses; female vocalizations were the result of a 'hormonal imbalance' or, like the horns of an antelope, simply a non-adaptive by-product of shared genetic architecture with the male.

'The accepted wisdom was if you hear a female sing, it's a functionless aberration – an old female that has a bit too much testosterone in her system,' Naomi Langmore, professor of evolutionary ecology at the Australian National University, told me. 'The textbook definition of birdsong is that it's "complex vocalizations by male birds during the breeding season". So it's actually defined as a male vocalization.'

This persistent all-male classification seriously ruffles Langmore's feathers. For the last thirty years she's been studying the complex vocalizations of female songbirds and fighting to get their voices heard. She's part of a pioneering group of scientists who, tired of dogmatic androcentric definitions of birdsong, took it upon themselves to trawl through all the available scientific data to demonstrate that, far from being dumb, 71 per cent of female songbirds sing.

What's more, their calls are worth listening to; what these avian divas have to say challenges fundamental assumptions about Darwin's theory of sexual selection.

According to Langmore, the century and a half of oversight isn't simply another casualty of old-school sexist bias. This prejudice is mostly geographic in origin. Songbirds, or passerines, are the largest avian order, comprising 60 per cent of all known birds. The defining features of the six-thousand-plus species are highly developed toes that enable them to perch and a muscular larynx-like structure – the syrinx – that bestows their vocal dexterity. Beyond that, evolution freestyled to create dainty tits, soaring swifts and frivolous birds of paradise, amongst myriad other forms. Together the 140 or so families represent one of the most diverse vertebrate orders, thanks to an explosive evolutionary radiation in recent geographical times that's enabled passerines to take over the world. The only continent where you won't hear birds sing is Antarctica.

Despite their global domination, songbirds have traditionally been studied in Europe and North America. These mostly migratory species are from a recently evolved suborder called the Passeridae in which females are, indeed, more vocally subdued. Those females that do sing, such as the European robin, tend to lack sexual dimorphism and so are easily mistaken for noisy males.

It's a totally different story in Langmore's native Australia and across the tropics. Had Darwin lived down under he would have heard dozens of bush and backyard species, from the chainsaw-mimicking lyrebirds to delicate fairy wrens, making just as much noise as the males.

'Working on birdsong as an Australian you're very aware of this misconception in the literature. I would read papers about birdsong being produced by males and then I'd go out and do fieldwork and be surrounded by female birds singing everywhere I looked. So I could hardly miss that anomaly,' Langmore said.

Songbirds are understood to have evolved in Australia some forty-seven million years ago. Given the prevalence of female birdsong in their place of origin, Langmore and co. wondered if they'd always sung. So they created an avian family tree to reconstruct the ancestral state and deduced that the earliest female songbirds were, indeed, a bunch of raucous divas.

'That really turned things on its head,' Langmore told me. 'For so long these ancient groups of Australian songbirds with singing females were considered oddballs. Now it's the northern hemisphere songbirds that turn out to be the oddballs.'

Langmore's discovery was indeed profound. It proved that female song wasn't some recent evolutionary kink, found only in the tropics. Female songbirds had always sung. What's changed is that, in some northern temperate regions, in the more recently evolved families of songbird, females have, for some reason, ceased singing. Which is a radically different evolutionary scenario than the framework proposed by Darwin.

'The question we should really be asking is not why do male birds sing, but why have some females subsequently lost song?' Langmore said to me.

Unlike male birdsong, research into female song is still in its infancy. But it appears that female songbirds use their vocal abilities primarily to compete with other females. They sing to defend their territories, breeding sites or mates from other females, or to lure males *away* from other females. This makes much more sense in hot countries like Australia, where breeding seasons are lengthy affairs and couples remain on their territory year round.

'There's a real role for the female to defend her territory because the male may die, he may divorce her, or be sneaking copulations next door with another female. In any of those circumstances, the female still has to be able to defend her territory against any intruders, and perhaps even sing to attract a new mate. So in these tropical regions, it's very valuable for a female to be able to sing,' Langmore said.

It's a very different scenario in the gardens of Europe or North America, where most of the songbirds migrate south in the winter. When they return the male generally arrives first and sings his head off to establish a territory and attract a mate. Females will shop around before choosing a male, which has, in some cases, resulted in elaboration of male song through sexual selection. The breeding season is short, however, so she has to get down to business and breed before she packs up to move south again. As a result she's less likely to be getting into fights with other females and selection for song has been relaxed.

Birdsong is clearly as adaptive for females as it is for males. Cunning experiments have even shown that migratory females that rarely sing, like American yellow warblers, can be induced to start singing if other females, albeit dummies, muscle in on their territory.

These female birds engaged in vocal combat over nest sites and territories pose a problem for Darwin's theory of sexual selection. This impasse has become something of a sticking point for evolutionary biologists.

'It's not that Darwin was wrong. To some degree, song elaboration in migratory males has still evolved through sexual selection. But it's just that that's only a small part of the story. It's not the whole story for all birdsong,' Langmore told me. 'What we're realizing now is that song has really got a much broader range of functions. It's involved in competition for all sorts of things. Not just for mates. So instead of thinking song has evolved just through sexual selection, we now think that song has evolved through social selection.'

The concept of social selection was developed by the theoretical biologist Mary Jane West-Eberhard back in 1979. West-Eberhard recognized that Darwin's theory of sexual selection was too narrow to explain elaborate traits that evolved as a result of competition, not for sex but for territories and resources outside of the breeding season, in females as well as males.

West-Eberhard did not attempt to discredit Darwin, but instead proposed expanding his principles so that sexual selection became a subset of the broader backdrop of social selection. She illustrated her argument, much like Darwin himself did, with a vast array of creatures whose flamboyant traits or sexual dimorphisms could not be explained by sexual selection alone, and might even have different social functions depending on the time of year and situation. She outlined how dung beetles' horns, pheasants' tails, toucans' beaks, birds' songs and the dominance behaviour of bees and wasps could all be explained by the broader category of social, if not sexual, selection.

The concept has been controversial nevertheless. Many zoologists don't see the need to invite yet another form of selection to the evolutionary party. Let alone one proposed by someone other than Darwin (and an American woman to boot). But as research into social competition beyond the eye-catching showmanship of males grows, so does an awareness that Darwin's definition of sexual selection isn't expansive enough to explain elaborate traits like birdsong or indeed bright plumage and ornamentation in female birds. Worse still, Darwin's narrow focus has 'clouded our view' and fostered a scientific bias that assumes such elaborate traits and sexual dimorphisms must all be to do with mating success, when they're often to do with other forms of social competition.

The debate is likely to rumble on. However we choose to label the forces driving these fabulous characteristics, it is becoming clear that females are just as competitive as males; it's just the focus of their efforts is often different. While males clash primarily over access to females, females are more likely to fight over resources related to fecundity and parenting. And while their efforts may be more surreptitious, female competition is just as influential in shaping the path of evolution as male fighting – perhaps even more.

MAKE WAY FOR THE ALPHA CHICKEN

When it comes to social species, status is key for determining access to food, shelter, top-quality sperm – all the resources a female requires to reproduce. So it pays to be the top bitch. Males may suck up all the attention with their bloody battles for supremacy, but group-living females generally inhabit some kind of hierarchy, often independent of the male order. The first dominance system ever fully documented was, in fact, feminine in nature. A young Norwegian scientist by the name of Thorleif Schjelderup-Ebbe introduced the scientific world to the very first alpha, which just happened to be a female chicken.

Schjelderup-Ebbe had harboured an almost obsessive interest in chickens since the age of six. This was the turn of the twentieth century, before young minds were bewitched by TikTok and Poké-mon, and the young Schjelderup-Ebbe began documenting the hens at his parents' summerhouse with such zeal he even travelled to visit them in wintertime in order to catch up on their social lives.

He noticed that during routine squabbles between pairs of hens in a group, one would peck the other. The pecker, generally the older of the two, would henceforth gain priority access over the loser to the best roosting sites and food. Following rounds of peck-ing contests, an overall champion would emerge and group aggression would cease, as each bird understood and accepted her place in the resulting hierarchy. The top-ranking hen, which he named the 'despot', would however routinely remind any subordi-nates daring to eat before her of their relative social status by administering a painful peck.

The young Schjelderup-Ebbe had discovered the original pecking order.

'Defence and aggression in the hen is accomplished with the beak,' he noted in his groundbreaking paper, '*Gallus domesticus* in its Daily Life', published in 1921.

Schjelderup-Ebbe had hit on something big. Bigger than just a bunch of back-biting hens. The young scientist hypothesized, quite correctly, that this sort of despotism is one of the fundamental principles of animal and bird societies. Unfortunately, an altercation with a more powerful female academic prevented him from receiving the kudos he deserved, suggesting he was better at uncovering hierarchies than navigating them in real life.

Schjelderup-Ebbe had recognized that female hierarchies were far from trivial. 'Fights among chickens, which are usually considered to be quite harmless, are certainly not so and do not result from a momentary whim,' he wrote. 'They put a lot at stake, sometimes even their lives, in order to win.'

This is true across animal societies, from birds to bees. The prize for clawing up the social ladder to alpha female status is a significant reproductive advantage, and it's worth fighting for. In males the battle for supremacy can be bloody and boisterous and hard to ignore. Female power struggles are generally far subtler, although no less devastating. Which is probably why many female hierarchies went largely unnoticed for decades.

'Females are not innately disposed to organize into hierarchies . . . primate males appear to be the archetypal "political animal",' was the woeful conclusion of *Female Hierarchies*, the first ever textbook specifically devoted to female dominance relations.

This could not be more wrong. Strategic competition amongst females is central to primate organization. Most female primate societies feature stable heritable matrilines, which compete with one another using psychological intimidation, tactical alliances and cruel punishment in their ruthless battle for control.

Recall the savannah baboons we met in the previous chapter. High-status females have it all: first dibs at food sources and a high-ranking protection racket for them and their babies. Low-status mums and their offspring are subject to constant bullying by those above. The resulting stress impacts their reproductive capacity. Low-ranking females breed later, ovulate less frequently and can even spontaneously abort as a result of being persistently terrorized by dominants.

As Sarah Blaffer Hrdy notes, across primate species 'high rank carries with it not only freedom from harassment and exploitation by more dominant females, but also this sinister prerogative to interfere in the reproduction of other females'.

This low blow to a female's fertility has grave consequences, arguably far greater than the most savage canine-baring contest between males. It hits a female where it really hurts, in her precious genetic legacy. Being prevented from breeding is the most ruinous punishment there is. Just because fists aren't flying 24/7, it's naive to assume female primates are not as competitive as males. They just fight craftier and dirtier.

A low-ranking female baboon's best chance of survival is to play a deft political game and ascend the social ladder through strategic coalitions that protect her and her offspring. Female primates have been described as 'obsessed with signs of status differences or disrespect'. They just don't make it as obvious as the males.

On the surface, a cluster of female baboons lounging around in the afternoon shade, snacking on seeds and checking each other's fur for ticks, may appear to be a picture of female harmony. Scratch the surface of this maternal haven and complex relationships are being calculated and negotiated by means of grooming, food sharing and babysitting. Individuals might work together to mob an aggressive male or even care for one another's babies, but they do so selfishly to protect their procreative potential. Such diplomatic manoeuvring requires significant cognitive power, and is likely to be one of the driving forces behind the increase in brain size and intelligence in all social primates, including us.

ALL HAIL THE REPRODUCTIVE AUTOCRATS!

Animal matriarchies are no feminist Eden. Often, an unpleasant undercurrent of reproductive tyranny manipulates the fine line between teamwork and exploitation. Nowhere is this so stark as in the collective lives of that loveable TV star, the meerkat (*Suricata suricatta*), whose violent totalitarian society is somewhat at odds with their saccharine screen image.

I'll admit, it's hard not to be charmed by a meerkat. I'm generally immune to conventional cuteness. I tend to gravitate towards strange slimy creatures with other-worldly lives. But when I visited the Kalahari Meerkat Project in South Africa a few years ago, I was completely won over by these small social mongooses. They're such manic comic characters. And their penchant for standing on their back legs makes anthropomorphizing an easy crime. Their default setting is to dig, which they do with great fury the moment they stop moving, generally to little or no avail. Scrapping over monster scorpions and toppling over whilst nodding off in the sun are further staples of a well-honed clowning repertoire. Behind the slapstick, however, meerkat society owes more to Stalin than it does to Chaplin.

Meerkats live in clans of three to fifty, with a single dominant female monopolizing 80 per cent of the breeding. The rest of the mob – her relatives, descendants and a few itinerant males – help out with territorial defence, sentinel duties, burrow maintenance, babysitting and even suckling the dominant's pups. This kind of division of labour, where only a few individuals get to breed and the rest of the group helps, is known scientifically as 'cooperative breeding'. This term has always struck me as wildly euphemistic. The meerkat's apparent camaraderie isn't achieved through cosy cooperation so much as outright tyranny.

Meerkat society is predicated on ruthless reproductive competition between closely related females who, when pregnant, will readily kill and eat each other's pups. This baby-eating bonanza is kept in check by the omnipotence of a dominant female with a zero-tolerance policy for breeding subordinates. Her goal is to prevent any of her female relatives reproducing during her reign, and rope them into caring for her babies instead. This removes any unwanted competition for her pups and protects them from being eaten. It also allows her to invest all her energy into raising more litters than she could do otherwise. It's a prize position that's worth fighting hard and dirty for. As the biggest and most violent meerkat in the mob, she'll use extortion, physical abuse, entrapment and murder to achieve this end.

Vacancies for dominance don't come up often. Generally the position only becomes available when a matriarch dies, perhaps at the talons of a hawk or a rival meerkat gang. The top job then falls to the oldest and heaviest female in the group, most likely one of the matriarch's daughters.

From the moment a meerkat inherits her supreme status her size increases, her testosterone levels rise and her hostility towards all other females will surge. She will demonstrate particular hostility towards those that are closest to her in age and size – most probably her sisters – and who therefore make up her greatest reproductive competition.

'If you are a female meerkat your best bet – your hope in life – is that someone will eat your mother,' Professor Tim Clutton-Brock, Cambridge behavioural ecologist and founder of the Kalahari Meerkat Project, explained to me over the phone in his cut-glass British accent and characteristic avuncular manner. 'And it's no good if they eat your mother at the wrong time. What you really want is for them to eat your mother when you are the oldest subordinate in the group. Otherwise one of your bloody sisters gets the job and she throws you out.'

Clutton-Brock has been documenting the savage soap opera of meerkat family life for twenty-five years. He explained that it's not just the matriarch's sisters that get booted out. During the dominant's reign any females that have reached sexual maturity, and might entertain the idea of motherhood, will be run out of the group before they can even try.

'You regularly see dominant meerkats evicting their older daughters. They are brutal: if their daughters don't get the hell out, they kill them. If you look at a meerkat group there are basically no subordinate females over four years old because the dominant females evict them between the ages of two and four. And so they're all gone.'

Eviction follows a well-worn programme of escalating abuse. Entry-level bullying begins with swiping snacks straight out of subordinates' mouths. It makes for an amusing vignette on TV, but the truth is darker. Edible life is scarce in the Kalahari. Most creatures that could be eaten by a meerkat have evolved noxious booby traps as protection: scorpions that inject lethal neurotoxins and beetles that shoot boiling acid from their anus at high speed. All require disarming before consumption. First, however, you have to find them. The baked earth of the desert can be as tough as concrete. It took me ten tough minutes with a pickaxe to experimentally excavate a scorpion after I'd located its den. Meerkats can only cope with quarrying soft sand and may churn through mountains of the stuff before finding something palatable. So having hard-won food stolen isn't just rude – it's a critical loss of expensive energy.

Next comes physical abuse: hip-slamming and the casual biting of tails, necks and genitals are all favoured moves by a dominant female looking to exert her power. Corporal bullying serves to impose authority and, as an added bonus, the resulting stress could also depress the victim's fertility. Its main purpose, however, is to make life so unpleasant for the victim that they will leave the group.

'Eviction starts with bullying. Bites at the base of tail are common. When you see a meerkat with a raw patch at the base of their tail you know that one is going to be next,' Clutton-Brock explained.

Exile may sound like a walk in the park compared to a constant barrage of snack swiping and genital biting. But the only thing more oppressive than a dominant female meerkat is the Kalahari Desert itself. Environments don't come much less forgiving than this vast semi-arid savannah. Rainfall is a hazy memory for most of the year and temperatures fluctuate by 45 degrees Celsius daily. During the height of summer the daytime temperature tops 60 degrees Celsius, yet in the winter the nights can be freezing. Without warm bodies to huddle next to in a communal den, a meerkat could easily go to bed and never wake up.

More deadly still is the risk of trespassing on a neighbouring meerkat mob's territory. Each group has a home range of somewhere between two and five square kilometres, which is vigorously patrolled and defended. Due to the scarcity of suitable burrows and food neighbouring groups are in intense competition and often engage in violent fights. According to Clutton-Brock his study area is one continuous patchwork of warring gangs. So an exiled meerkat cannot avoid entering a rival's neighbourhood. As soon as she's spotted by the resident mob, they'll chase her away. If they catch her, she's dead.

If a rival gang of meerkats doesn't kill an exiled female, there are dozens of sharp-eyed predators only too eager to turn a lonely meerkat into dinner. The soft sand that meerkats need to forage in is only found along dry river beds, grasslands and dunes with little or no vegetation for cover. That leaves a hungry meerkat highly exposed – out in the wide open, digging away with their head in the sand. Without a sentinel to keep watch and sound the alarm for predators, a solitary meerkat is easy picking for umpteen aerial predators, wild cats or jackals.

It is the brutality of the Kalahari that gives a reproductive auto-crat the necessary leverage to enforce her totalitarian regime. Solo survival is a specialist sport and for many animals it's work together or die. Add to this the fact that the Kalahari is an ancient desert, some sixty million years old, and you have the perfect conditions for the evolution of some seriously twisted 'cooperative' relation-ships. As well as the meerkats there are ants, termites, colonial birds like the pied babbler and vast burrowing societies of Damaraland mole rats. All have converged on a version of reproductive totali-tarianism and associated 'cooperation' in order to survive.

As the travel writer A. A. Gill once noted, 'There is no romance here. The Kalahari is an amoral, unregulated market force, a pure vicious capitalism practised by professionals.'

In the event that a developing female meerkat's hormones override the system, and she has the temerity to get knocked up by a roving male, retaliation is swift and terminal. The pregnant subordinate will be unceremoniously evicted. The ensuing stress generally trig-gers her to abort. If she manages to go full term without detection and give birth in the den, the matriarch will kill and eat any unwel-come pups – very often her own grandchildren – and banish the female from the group.

If that wasn't chilling enough, there is an additional 'coopera-tive' twist to the eviction of these recently bereaved daughters. They may be permitted to slink back to the group on one condi-tion: they take up wet-nursing duties for their murderous mother's babies.

Suckling is a serious drain on a subordinate's reserves, but these enslaved females have no choice when the alternative is exile and a lonely death. This threat explains the curious altruism of inher-ently self-motivated individuals. Taking care of the dominant's pups is a form of punishment or 'rent' to be paid for their errant behaviour. Given the close relatedness of the females in the group, helping raise their mother's offspring does at least mean they share a significant part of their genome. This genetic connection strength-ens a subordinate's incentive to sacrifice and offers some hereditary benefit to cooperation. If a subordinate female manages to appease the dominant and remain in the group for long enough, after all, there is always the chance she may one day inherit the title and get to breed herself.

'As a developing female, the last thing you want is for your mother to kick you out of the group. So in a sense, you have to play her game. She is bigger than you and has the capacity to throw you out. So instead you just have to hope someone will eat her,' Clutton-Brock told me.

One thing you won't see is subordinate females ganging up to overthrow the top bitch. 'That's what a primate would do. Have an alliance and take on the dominant,' Clutton-Brock explained. 'Meerkats don't form alliances. They're very stupid – they're cer-tainly not what you'd want to rely on to decide your insurance policy.'

Insurrections only occur if dominance is ambiguous. Say if a weak female inherits the top job, by virtue of being oldest but perhaps not the heaviest. Alternatively, a strong dominant might get sick or injured. The system then breaks down and bloody chaos ensues, as closely matched females make a desperate play for the top. In addition to escalating aggression towards one another, they'll often all fall pregnant. The result is a baby-eating blood-bath. The first litter born will be eaten by a pregnant female, then the next. This massacre continues until the last litter is born, which will be the only one to survive regardless of who their mother is. In one study, of 248 recorded litters 106 failed to emerge from the den, suggesting their pups were all killed.

Meerkat culture is tense and homicidal. A study investigating lethal violence in more than one thousand different mammals unmasked the meerkat as the most murderous mammal on the planet – beating even humans to the brutal top spot. Every meerkat born has a one in five probability of being killed by another meer-kat, most likely a female and quite possibly their own mother.

All of which makes the meerkat an odd choice for wholesome family entertainment or as a trusted purveyor of car insurance. They may be cute and comical and their society labelled 'coopera-tive', but each individual is nevertheless out for themselves. No god would have created such a flawed and blood-stained system. But evolution did and, somehow, it not only works, but is highly suc-cessful. Dominant females can manage three to four litters a year, instead of just the one which might be expected for an animal their size. One legendary matriarch, named Mabili, succeeded in birth-ing eighty-one pups during a decade-long reign.

Given that only one in every six or seven meerkats will get to reproduce effectively, the variance in breeding success for female meerkats is even more profound than for your average antagonistic alpha male. Clutton-Brock told me that the most successful red deer stag he ever documented only ever fathered about twenty-five offspring during its lifetime, despite all the effort of growing enor-mous horns, fighting off rivals and energetic procurement of a large harem.

LONG LIVE THE QUEEN!

Cooperative-breeding females wipe the floor with that figure. The queens of the co-op life are, of course, the social insects – namely all species of termite and ant, as well as some wasp and bee – whose societies are a wonder of procreative totalitarianism. Only one in tens of thousands of females ever get the chance to become a mother. Those that do can be extraordinarily productive since their sole job is to lay eggs. This they do in the safety of their very own royal chamber aided and abetted by supporting castes of sterile workers and soldiers.

This system has been honed to the extreme by the termites, who have been cooperating since the early Jurassic, some 150 million years ago, when dinosaurs walked the earth. Those belonging to the species *Macrotermes bellicosus* not only farm fungus but also create towering humidity- and temperature-controlled mounds up to nine metres tall on the savannahs of West Africa. At their centre is the royal chamber. Unlike with ants and bees, this includes both a queen *and* a king, who mates with her for life. In many species the queen swells into a monstrous egg-laying machine whose abdomen has swollen over a thousand times into a giant waxy off-white sausage around ten centimetres long. Her head, thorax and legs remain tiny and can only flail about pathetically, since all other movement is restrained by her grotesque, pulsating girth. She must be fed and her gargantuan maggot-like body cleaned by a legion of workers, allowing her to spend every bit of her energy squeezing out a fresh egg every three or so seconds, all day every day, for up to twenty years. At over twenty thousand eggs a day, she's capable, in theory, of producing some 146 million termites in her lifetime, making her the most reproductively successful terrestrial animal on the planet.

This extreme brand of cooperative breeding, involving such a clear division of reproductive labour between breeders and infertile working castes, is known as eusociality – from the Greek eu- meaning 'good'. Although this is another highly subjective term since, in

truth, it is only really 'good' for one individual: her Royal Reproduc-
tiveness. The rest of the several million termites in the colony, other
than the king, are rendered sterile and kept in their lowly castes by
ingesting pheromones secreted by the royal anus, all of which makes
the British monarchy suddenly seem quite reasonable.

Eusociality is an alien way of life that challenges philosophical
ideas of the individual and has provided inspiration for countless
sci-fi dystopias. Aldous Huxley reportedly based the dictatorship
in *Brave New World* on humans as social insects, with five castes.
The fact this sci-fi society has evolved in invertebrates – such distant relatives of humans – is no doubt reassuring. There is, however,
one mammal society that's been classed as eusocial: the fantasti-
cally weird naked mole rat.

Meeting a naked mole rat has been top of my animal bucket list
since, well, for ever. But the world's only eusocial mammal isn't
exactly easy to find. For a start they spend their entire life underground, where they live, much like termites, in colonies of up to
three hundred in vast tunnels, some several kilometres long, under
the arid grasslands of Ethiopia, Somalia and Kenya. Their dark,
claustrophobic existence means that evolution has done away with
unnecessary extravagances like eyesight and fur, which are surplus
to requirements for the task at hand, namely the relentless quarrying for edible plant tubers. These are hard to find under the deserts
of East Africa and the driving force behind the mole rats' need for
collaboration – a colony can shift 4.4 tonnes of soil a year in the
desperate search for food.

Mole rats rarely, if ever, venture on to the surface, which is perhaps wise. Like the Kalahari, this corner of sub-Saharan Africa is
highly inhospitable. The fierce equatorial sun would roast a naked
mole rat in a matter of minutes. It's also a hotspot for banditry and
terrorism. Less an issue for mole rats than it is for their researchers.
I'm told that at least one mole rat scientist 'went missing' in the
field, and has never been found.

Every time I've visited Kenya I've searched for the volcano-like
molehills amongst the baked red earth – the only above-ground

evidence of their clandestine labyrinth – with no joy. When I eventually encountered my first colony, it wasn't in the African badlands but a very hot cupboard in East London. Dr Chris Faulkes, a leading authority on mole rats, maintains a study colony at Queen Mary College, University of London. It turns out they've been there for almost thirty years on the top floor of the zoology department – a mere stone's throw from my home.

My long-awaited naked mole rat liaison wins the prize for being the most surreal animal encounter I've ever had. COVID-19 was raging in London and Faulkes met me wearing a mole rat facemask, before bounding up seven flights of stairs and ushering me into a small stuffy room that smelt of sweet yeast. There he'd created a facsimile of the mole rats' subterranean desert life using Tupperware boxes and oodles of clear plastic tubing held together with gaffer tape, arranged over half a dozen wide shelves. It was a decidedly Heath Robinson affair, which nevertheless allowed Faulkes and his research team to observe the animal's otherwise secret social life in an attempt to decode it.

Scurrying along the tubes, like some sort of factory production line, were what looked like hundreds of uncooked sausages on legs. Naked mole rats are truly exceptional-looking creatures that regularly top the ugly animal charts. Their Latin name, *Heterocephalus glaber*, means animal with a 'loose skin' and an 'oddly shaped head', which merely hints at the idiosyncrasy of their appearance. 'They look like a penis with teeth,' Faulkes announced with surprising candour from behind his mask bearing the same image, before adding, 'I think they're very cute.'

The naked mole rat's face is one only a mother could love – if only the mother in question were not a highly belligerent despotic queen with no love for anything. Their wrinkly pink body is indeed remarkably phallic, and protruding from the tip of a helmet-like head are two pairs of terrifyingly long yellow teeth. These are especially prominent since the mole rat's lips close behind their ever-growing gnashers so they don't end up choking on soil while they use them to dig. With little more than two black dots representing their useless eyes and no external ears, the overall effect is indeed the-Johnson-from-hell, which is somewhat ironic for such a female-dominant species.

Faulkes plunged his hand into one of the lunchboxes, fished out a sabre-toothed sausage by its stumpy bald tail and handed it to me. The mole rat's nude skin felt stretchy and unbelievably soft. Faulkes explained that this helps reduce damage from the abrasive nature of tunnel life. Naked mole rats produce a novel type of hyaluronan – the same interstitial gloop you'll find in expensive face creams promising eternal youth. It makes their skin extra-specially elastic and as an added bonus it might also be the reason they don't get cancer.

Naked mole rats are a true scientific wonder. The world's only cold-blooded mammal is apparently immune to cancer, capable of surviving eighteen minutes without oxygen and feels no pain. These almost indestructible rodents can live for over thirty years – eight times longer than would be expected for an animal their size. All of which has made them of particular interest to life-hacking labs in Silicon Valley searching for the fountain of youth, but in reality, these are all otherwise unenviable adaptations to an extreme sub-terranean life.

Faulkes has spent the last thirty years investigating the naked mole rat's eusocial society. Colonies are ruled by a single queen who does all of the breeding with one to three selected males. Four to five times a year she'll give birth to around a dozen pups, although she can squeeze out many more – an eye-watering twenty-seven was recorded in one litter. Even though pups are born decidedly fetal – bright-red jellybeans with transparent 'gelatinous' skin – this is an extraordinary output. In order to pump out this many babies, the queen's body can become grossly distended, just like a termite's. Like a social insect queen she also lives ten times longer than the workers of the colony, but with no age-related decline in her fertility, thereby allowing her to leave an extraordinary genetic legacy during her abnormally long reproductive life. One legend-ary queen reared more than nine hundred pups during twenty-four documented years.

Other than the queen's chosen mate, the rest of the colony act as workers or soldiers. Unlike in social insects, Faulkes tells me these subordinate roles aren't hardwired, they're plastic. The bigger mole rats tend to take on the soldier's role of defence against invading foreign mole rats or predators like snakes, while the smaller ones are the workers and spend their days digging for tubers, sweeping the tunnels with their bristly little back legs, tending to babies or cleaning the toilet chamber. Mole rats are, it seems, unusually fastidious about their waste. According to Faulkes they have personalities and some seem to *prefer* certain duties. This is just as well, since unless they rise to queen or top male breeder, these individuals will sacrifice their whole lives just to keep the colony going.

'99.99 per cent of the colony will never reproduce,' Faulkes told me. Unlike meerkats, subordinates never flout the no-breeding rule with sneaky copulations. They can't. The queen has put paid to any notion of parenthood by suppressing their sexual development. Both male and female subordinates remain trapped in a pre-pubescent state. They don't even develop adult genitalia. So subordinate sex is ruled out in this eusocial system. As a result, Faulkes tells me, the non-breeding castes are impossible to tell apart sexually. They scuttle about the colony like unisex smoothies doing the queen's bidding.

Faulkes has spent his entire career trying to figure out how the queen achieves this feat of sexual suspension. Originally it was thought that, like social insects, pheromones were the answer, which the queen sprinkled around the colony in her urine like bromide in a soldier's tea. Faulkes proved that wasn't the case. Other researchers have suggested that subordinates are manipulated by eating the queen's chemically controlled faeces. Mole rats are enthusiastically coprophagous characters. Baby mole rats routinely beg faeces from their carers. Eating excrement replenishes valuable gut bacteria, nutrients and water, after all. But Faulkes explained that for the queen to use faecal matter for such widespread colonial control would mean 'shooting it out like a machine gun', which Faulkes has yet to witness.

It is Faulkes' belief that the queen keeps her subordinates imprisoned in their sexless state by low-level physical bullying. She is often seen to be giving workers 'shoves' to let them know she is still a strong and capable leader. To avoid attacks, workers and soldiers must be submissive; if a subordinate meets the queen in one of the many tunnels of the colony the queen will pass over and the subordinate will be left to squeeze under her.

The exact underlying mechanisms of this physiological suppression are still to be worked out, but on a very basic level, Faulkes believes that these antagonistic cues from the queen translate into a state of 'stress', which alters the subordinate's brain chemistry and impinges on the hypothalamus, the area of the brain responsible for controlling reproduction. It's Faulkes' belief that the hormone prolactin plays a key role.

'What we've found is that non-breeding male and female mole rats have really high levels of prolactin, which in humans is kind of instant infertility,' he told me. Prolactin is high in pregnant women and new mothers – it stimulates the breasts to make milk and reduces fertility until the current offspring is weaned. It's also associated with parental care. So it is possible that there is a dual role of prolactin in mole rats: it not only stops them from breeding, but also makes them better carers to the colony's young.

For the queen to maintain her dominance throughout the colony takes some serious legwork.

'We know she expends enormous amounts of energy patrolling the colony. We found that the queen is more than twice as active as the next most active animal in the colony and travelled three times the distance over the space of about eighteen months,' Faulkes explained to me.

Faulkes thinks this relentless royal tour is necessary to maintain the colony's sexual suppression. 'The queen is really not like a lazy monarch, just slopping around in the nest chamber and being fed by the others [like a social insect],' Faulkes told me. 'It's really quite a tough position to be in because she has to maintain her dominance the whole time.'

Colonies can function for years, if not decades, in a state of peaceful industry as long as the queen is a dominant, mobile presence. But if she is weakened or removed for any reason, all hell breaks loose. Her absence will trigger the next highest-ranking females to become sexually mature within a week and then, very rapidly, it all turns very *Game of Thrones*.

' "When you play the game of thrones you win or you die" – that's the famous quote and it really does apply to naked mole rats. Because they're competing for the throne and they'll kill anyone or die in the process. It's super brutal,' Faulkes said.

These professional burrowers have evolved a ferocious bite. A quarter of their entire body musculature is devoted to powering their jaws to crunch through baked earth – and then there's those javelin teeth. When females turn these industrial excavation tools into weapons things get ugly fast. Faulkes told me the plastic tubes of the colony become coated in blood and his team have to try to separate the warring females. 'You have to be bloody careful when it's all of these gnashing teeth of death – they'll take your finger off.'

Witnessing a pair of wrinkly pink penises savage each other to death with long yellow teeth is disturbing in many ways. 'It's horrible when they do it, because you feel a bit powerless. We always get upset, especially when you get a phone call on Saturday morning from the animal person who's covering the lab for the weekend. "Oh, they're fighting. I don't know what to do." '

The extreme violence is down to the fact that, for the females involved, the vacancy offered by a weak or missing queen offers a once-in-a-lifetime opportunity to breed. Striking out alone is simply not an option in this netherworld. Naked mole rats can only exist as a self-sacrificing multitude. They've evolved to thrive in the most inhospitable environment imaginable by banding together, divvying up the tricky job of survival and sharing the load. They are an incredible advert for the power of cooperation, but also reproductive despotism.

These outbreaks of extreme female violence, amongst an

otherwise quietly successful culture, are a reminder of the intense drive for a female to reproduce, and the simmering reproductive competition that underpins eusocial societies, and necessitates the most oppressive leadership found in the animal kingdom.

As Faulkes said, 'Naked mole rats are a brilliant example of a sort of utopian, almost communist society, but of course, lurking under that, there's all kinds of sinister shit going on.'

CHAPTER EIGHT

Primate politics: the power of the sisterhood

In the movie *Madagascar*, the great African island is ruled by a fast-talking ring-tailed lemur called King Julien XIII. Blockbuster animations are not famed for their realism, but given King Julien is from a species that actually calls Madagascar home you'd be forgiven for thinking he's one of the more credible characters in the film. In fact, he's just as implausible as the movie's misplaced penguins. In real-life Madagascar there are ring-tailed lemurs aplenty, but their leader isn't a king, she's a queen. The producers may have felt it only natural to impose male governance on their feel-good movie, but in ring-tailed lemur society females are unquestionably the authoritarian sex.

It's the same story with most lemur species. This peculiar group of primates, found only on the island of Madagascar, are largely female dominant. Whether they're in monogamous partnerships, like the eerie singing Indri (*Indri indrii*), or polygamous societies, like the black-and-white ruffed lemur we met earlier, in 90 per cent of the III species it is the females who call the shots sexually, socially and politically. Madagascar is an island of bossy bitches – a land where female primates rule.

This tale of female empowerment provides sharp contrast to the familiar patriarchal narrative of primate life, featuring a despotic male chimp beating his chest and terrorizing society with brute force – a popular social model for our own ancestry. Such male dominance is considered a basic prediction of Darwin's theory of sexual selection – a by-product of male competition over females, which selects for big, aggressive, weaponized males who can then take advantage of their physical fortes to subjugate smaller, passive females.

As a result, female dominance has long been considered rare amongst mammals. The spotted hyena and mole rat matriarchs we've encountered in the book so far have evolved a significant size advantage, which allows them to reverse Darwin's 'natural order' and overpower males. Lemurs show little in the way of sexual size dimorphism, however. So how has the weaker sex become the ruling class? And what can lemur society teach us about the origins and dynamics of power in other primates – including ourselves? I made a pilgrimage to the scorched hinterland of southern Madagascar to find out.

Madagascar is the second largest island country on the planet and, despite rich natural resources, one of the poorest – three quarters of the population survive on less than two dollars a day. These facts conspire to make travelling there ill-advised for anyone in a hurry. On a map, my journey from the coastal town of Morondava to the Ankoatsifaka Research Station in Kirindy Mitea National Park, in the country's remote and arid south-west, looked to be little more than forty kilometres. As a seasoned Africa traveller, I assumed it should take around two hours. This proved to be woefully optimistic. Once we left the sleepy coastal town the drive morphed into an extreme episode of *Top Gear*.

Randzi, my driver, remained as cool as a cucumber as we skidded along the endless fine white sandy river beds standing in for roads. Occasionally he would abandon our four-wheel drive without explanation – we shared only a few words of French and our accents were quite incompatible – and head off into the blinding heat haze on foot to check for spots of sinking sand. The only other traffic we saw was the occasional dilapidated wooden cart, pulled by pairs of heavily beaten and belligerent zebu. We never got above second gear – our snail's pace further slowed by umpteen 'tolls'. These rudimentary roadblocks were manned by enterprising local Vezo tribespeople, the women's faces covered in creamy yellow tree bark pulp to protect them from the blistering sun, demanding meagre amounts of money to let us pass.

It was slow progress, but I felt exhilarated by the adventure. That was, until we lost our way and I temporarily misplaced my *bonne humeur*. I'd been warned against driving at night, not because of the treacherous 'roads', but the threat of bandits. Cattle rustling is a way of life in these parts – banditry being one of the few career paths available in this sparsely populated semi-desert. We had no mobile phone signal, no map, no one to ask and an extremely out of the way destination, which, on account of the language issue, I wasn't entirely certain Randzi even knew.

I had no water left and just as I thought I might have to start collecting my tears for emergency rehydration purposes, we caught site of the field station at the end of a long, red sandy track, surrounded by a dusty and decidedly dead-looking forest. It wasn't much – just a small wooden shack that acted as a kitchen, and a lean-to that functioned as pretty much everything else – but it was more habitation than we'd seen in hours. The giveaway was the presence of an attractive woman in her forties wearing the kind of utilitarian clothing that screams field scientist. This was my host, Dr Rebecca Lewis, associate professor of biological anthropology at the University of Texas, Austin, and a leading expert on female dominance in lemurs.

Lewis helped me pitch my tent in a small clearing before whisking me deeper into the forest to catch sight of her subjects before they went to sleep for the night. As we crunched through the desiccated leaves littering our path, I felt rising flutters of excitement. Lewis studies white sifaka, *Propithecus verreauxi*, which just happened to be in my top ten of must-see species. Long before I knew they were female-dominated I wanted to meet these strange bug-eyed, snow-white prosimians because of the way they move: sifaka don't walk, they dance.

The deciduous forest the sifaka call home is made up of some of the most tenacious plant life on the planet, able to withstand months without rain – a world away from the majestic fecundity of the rainforest. In their place grows a grey confusion of spindly trees with branches that aren't strong enough to support this lemur's weight – roughly that of a cat. This makes travelling quadrupedally along branches or swinging from tree to tree like their monkey cousins out of the question.

Sifakas have got around the problem by evolving oversized feet and hands to wrap around and cling to the trunks of the trees, and long legs for leaping. They do a brilliant job – the sifakas ping about like balls in a pinball machine as their powerful elongated thighs propel them up to thirty feet from one trunk to the next. Their locomotive system only exposes flaws when a sifaka is forced to the ground where their long legs, short arms and giant comedy feet make walking on all fours impossible. Instead they must bounce sideways, arms outstretched for balance. It's a great example of why evolution is true – no god would have designed such a crazy means of moving, unless they had a truly wicked sense of humour.

'To me they're like Wonder Woman – they're brilliant leapers and they have all the power,' Lewis says of her lemurs.

By the time we caught up with the sifaka they'd clocked off leaping and dancing for the day and settled into some serious lounging. Nevertheless, there were hints of female ascendency even in repose. Sifaka live in small family groups of around two to twelve individuals generally comprising a matriarch, her offspring and one or two adult males. The matriarch of the troop I met, known as Emily to Lewis, was snuggled with her young at the top of the tree in a cosy huddle, in preparation for the chilly night ahead. Meanwhile Mafia, the adult male, was sitting below them – an oft-seen physical demonstration of rank – with no other body to cuddle than his own. Temperatures can drop to a brisk ten degrees Celsius at night and Lewis told me she frequently sees males left out in the cold like this.

Male sifaka are second-class citizens, forced to give up the comfiest, sunniest sleep spots and best food to the alpha female. Any resistance is met with a firm hand. According to Lewis, I'd arrived at the best time of year to see the females exercising this kind of authority. During these parched winter months, most of the trees have shed their leaves, reducing Kirindy to a barren desert of woody skeletons. I'd certainly never experienced a tropical forest so devoid of life, and so eerily quiet – no chirping insects, no birdsong, just the crackle of footsteps on dead leaves to break the silence. There was clearly little for a leaf-eating lemur, or indeed any animal, to eat. 'The sifakas lose 15 to 20 per cent of their body mass in winter. It's a really hard time,' Lewis told me.

Baobab trees offer the sifakas an oasis of sustenance. These goliaths of the rooted world store water in their fat barrel trunks and, when the rest of the forest is all but deceased, they produce their fruit: green velvet balls the size of oranges containing lipid-rich, high-calorie seeds that dangle like Christmas baubles from the baobab's stunted branches. The only problem is that the fruit shell is hard and sifaka teeth are not. Their incisors have fused into a delicate toothcomb adapted for grooming soft fur, not cracking wood.

'The male will spend for ever gouging and gouging to get through this woody shell to the oily seeds inside,' Lewis told me. 'They spend for ever damaging their fragile toothcomb and when they finally get in the female goes *whap*, hits him across the head and says, "Thank you very much. I'll take that!"'

I caught the tail end of one such interaction the following morning. We headed into the forest at around 9 a.m. to catch the sifakas while still in their sleep tree. It was a surprisingly late start – not the dawn patrol I'd experienced studying other primates. Sifakas are slow to get moving after a cold night in the forest. After a spot of sunbathing, they bounded off into the tinder thicket in search of breakfast, with Emily leading the way. The sifakas moved through the tangled forest much quicker than we did and by the time we caught up with them in the baobab I could hear loud squabbling from the lofty branches and the unmistakable *chittering* sound of male submission. Next thing I knew, Mafia, the adult male, was pogoing along the ground, picking through the leaf litter looking for dollops of bright-orange baobab flesh, seeds still attached, the discarded remains of the feast taking place above him. Lewis told me this was quite common. After a male has had several baobab fruits stolen from him and endured enough beating, if he's smart he'll retreat to the ground in search of leftovers.

'I don't know why the males stick around, to be honest,' she said. 'They get beat up left and right, they don't get the good food. It's not an easy life.'

The aggressive dominance of the female lemur has troubled scientists ever since it was first discovered in the 1960s by a young American-born scientist named Alison Jolly. Jolly, who died in England in 2014 aged seventy-six, is one of primatology's lesser-known female visionaries. She pioneered a brand of environmental activism that helped protect much of Madagascar's unique wildlife and established the idea that primate higher intelligence evolved to manage complex social relationships rather than toolmaking. This flew against the thinking of the time but is taken for granted today.

Jolly was the author of over one hundred scientific papers, yet, despite such academic achievements, she was overshadowed by her contemporaries – Dian Fossey and Jane Goodall – and her contributions to science have somehow gone under the radar. Perhaps that was down to the heretical nature of her research. For while Fossey and Goodall were describing dominant silverback gorillas and the hierarchies of male chimpanzees on mainland Africa, Jolly was in Madagascar, quietly documenting something completely different – antagonistic alpha females.

Jolly arrived in Madagascar in 1962, aged twenty-five, with a brand new PhD from Yale and a big juicy 'Sputnik-era research grant to swell my pride'. She established herself at Berenty, in the isolated south, and set to work documenting the island's peculiar and poorly understood primate life. Her chief obsession was the absurdly charismatic and now justifiably famous ring-tailed lemur,

Lemur catta. Jolly discovered her stripy superstars engaging in some quite astonishing behaviours previously considered to be the exclusive preserve of masculine privilege.

For a start, it is the females that do the majority of territorial defence. They have well-developed scent glands and produce more chemical signals than the males – the reverse of what you'd expect. They seem to be more interested in the scent of their own sex than males, especially breeding females. The healthiest females produce a lot of fatty acid esters, which are a signal they are strong and sexy. Again, this is something normally only done by males. It suggests their scented signals are probably linked to competition with other females, while males don't pose much of a threat to them and can safely be ignored.

Females leave more marks in 'confrontation zones' and during battles between troops. When they stray into neighbouring territory, they keep sniffing the scent marks of the neighbours but don't deposit their own.

This is interesting, because it resembles what happens when male chimps go on 'patrol'. They get all hyped up but when they leave their own territory and go into the neighbours' they suddenly become very quiet because they don't want to get caught. If they do come across foreigners, they make a huge fuss, screaming and drumming on the trees. These female lemurs are doing something very similar but with scents. By sneaking next door and checking out the smell of the neighbouring females, they can size up the competition, but by not leaving any of their own, they can do it surreptitiously, avoiding retaliation from the neighbours. If they *do* encounter anyone, they suddenly start scent-marking like mad, trying to scare off the others. It's rather subtler than screaming and banging on trees but it's essentially the same thing.

Female ring-tailed lemurs do get physical. They've been described as 'exceptionally aggressive' towards *both* sexes. They will terrorize and even evict subordinate females, which for a group-living species can be a death sentence. They show no mercy to mothers carrying babies, which often get killed in the crossfire.

Female hostility over rank isn't unusual amongst primates – we've already witnessed bullying at play within the matriarchy of savannah baboons – but attacking males is. One study into female ring-tailed lemur aggression found that males are three times more likely than females to be the recipients of serious injury. Some males even die as a result of female violence.

In Madagascar I watched male ring-tailed lemurs subjected to routine physical harassment – biting, shoving and hitting – to surrender their food, a cosy sleep spot or prime patch of sun. Like the sifakas, these stripy lemurs take their sunbathing very seriously – legs akimbo, arms outstretched and eyes rolling back in ecstasy as they soak up rays with the unashamed enthusiasm of a Brit in 1970s Benidorm. If a male dares to occupy the sunniest spot during their pre-breakfast bake, he's swiftly and forcibly evicted.

Jolly noted how male ring-tailed lemurs are 'afraid of the females' on account of their 'frightening' manner. Females use their power for their own gain but also, on occasion, to maintain order. Jolly described an incident where a female intervened to stop a juvenile being bullied by an adult male, and 'put him in his place', and concluded that ring-tailed lemurs are the only wild primates 'where all females can be said to be dominant over all males'.

Jolly's radical observations were published in 1966 in a book innocently entitled *Lemur Behaviour: A Madagascar Field Study*. In addition to cataloguing her meticulous observations of feminine supremacy, Jolly also dared to suggest that her badass female lemurs might also provide 'a particularly exciting glimpse of our history'.

Madagascar's lemurs are prosimians – our most basic primate cousins. They evolved from an early offshoot of the main primate evolutionary path that subsequently went on to divide into the New World monkeys (simians that inhabit the Americas) and Old World monkeys (simians of Africa and Asia, which led to the evolution of all great apes, including us). Some fifty to sixty million years ago the ancestors of modern lemurs became isolated on the island of Madagascar. No one is quite sure how, but the most popular theory is they floated to the island on a raft of vegetation. This pioneering group of ancestral primates then evolved in isolation on this vast and relatively unpopulated island, radiating into a remarkable diversity of species from the tiny mouse lemur, *Microcebus berthae*, weighing no more than thirty paperclips (it is the world's smallest primate), to one the size of a silverback gorilla, *Archaeoindris fontoynontii* (now sadly extinct).

Jolly felt that where these three evolutionary lines – the lemur, the New World monkey and the Old World monkey – converged offered valuable insights into our shared ancestor, 'the not-quite-monkey who first formed social bonds with others of its kind'. Her discovery of this early branch of fierce, frightening females eroded the idea that aggressive male patriarchy is the natural state of affairs for all primates. Or, at least, it should have.

Jolly's groundbreaking discovery fell on deaf ears. Even her most studious revelations failed to undermine the prevailing idea that 'order within most primate groups is maintained by a hierarchy, which depends ultimately primarily on the power of males'.

In the sixties and seventies primatology was well and truly hyp-notized by showy male dominance systems. This obsession began way back when the science emerged in the 1920s. The zoologist Solly Zuckerman's pioneering work on baboons (part of the Old World monkey line) set the tone. 'Female baboons are always domi-nated by their males, and in many situations the attitude of a female is of extreme passivity,' he wrote in 1932. Although Zucker-man's colony was captive, overcrowded and unrepresentative of the wild, his observations grew into a theory that was to become a hallmark of primatology: the male dominance hierarchy is the defining principle of primate life. It governs access to resources (namely food and those 'passive' females) and is established by the ability to fight.

Following World War II, a preoccupation with the origins of human warfare quickly hijacked the emerging science of primatology, just as it was gaining momentum. Baboons, genus *Papio*, became the go-to model because they live in large

social, semi-terrestrial troops on savannahs, similar to the environment scientists believed our ancestors inhabited. Their brutal
culture suited the perceived importance of male dominance and
aggression in our ancestry. Male baboons certainly are intimidating; they can be up to twice the size of females and have evolved
terrifying canines, as long as a leopard's, in order to compete for
control of a harem.

Later, in the late seventies, chimpanzees took over as the model
for human ancestry. Jane Goodall's revelations of their warlike
nature fuelled the idea that human males must be pre-programmed
for violent supremacy, an idea made popular by the likes of Richard Wrangham, Harvard professor of biological anthropology and
one of many influential male scientists to promote our primate
ancestors as mirror images of chimps: patriarchal, male-bonded
and highly antagonistic.

'The search for our ancestry pulls out at last an image terribly
familiar and dauntingly similar to something we know in the contemporary world: a modern, living breathing chimpanzee,' Wrangham
wrote in his unambiguously titled bestseller *Demonic Males: Apes
and the Origins of Human Violence*.

This enthralment with a handful of terrestrial, Old World species
as models for human evolution resulted in what the anthropologist
Karen Strier has dubbed 'the myth of the "typical" primate'. These
idiosyncratic macho monkey societies became the perceived blueprint for *all* primates. Subsequent phylogenetic research has, however,
revealed that Old World monkeys make for poor primate prototypes.
Their behaviour is actually highly derived, tailored to meet specific
environmental challenges and far from representative. Primate societies in general are way more diverse than the familiar patriarchal
model of baboons and chimps. But this natural diversity was
overlooked – not just the lemurs but also the New World monkeys.*

---

* A paper by C. H. Southwick and R. B. Smith, published in 1986, demonstrated
that in the fifty-year period from 1931 to 1981 only ten genera account for over
60 per cent of all publications based on primate field studies. All but one of these
genera are from the Old World.

New World monkeys split from Old World monkeys some forty million years ago. They inhabit Central and South America and, as among the lemurs, antagonistic male dominance is unusual. Most species are peaceful and egalitarian, like the owl monkey we met earlier in which the sexes are the same size and parental duties are shared. Where dominance is displayed, as in the tiny (and super-cute) squirrel monkeys, marmosets and tamarins, it is the females which appear to have the upper hand.

As Lewis rightly proclaimed after dinner one night at our campsite, 'People think lemurs are weird. Well, New World monkeys are weird too!'

Back in the 1960s Jolly's feisty female lemurs certainly didn't fit with the popular testosterone-fuelled model of Old World primate behaviour. Their dominance over males, rather than providing a fascinating insight into the social evolution of our primate ancestors, was either ignored or became bogged down in semantic debate. The reluctance to recognize female dominance saw it swept aside as strategic male 'chivalry' or demoted to simply female 'feeding priority'.

According to Lewis, the study of female dominance remains 'intellectually isolated', often dismissed as 'a weird quirk of Madagascar' – the ultimate untestable hypothesis. But given that 10 per cent of lemurs are not female dominant, this decidedly unscientific argument doesn't hold a lot of water. It also ignores the global range of mammals, from carnivores to rodents and hyraxes, that have also been described as female-dominant, suggesting it can't just be a Madagascan phenomenon.

What is intriguing about female lemurs is they dominate without being physically more powerful. Aside from a handful of species in which females are marginally bigger, most are monomorphic – males and females are the same size – which is generally associated with a more egalitarian society. So how do female lemurs manage to get their way without the brawn to back up a threat?

Lewis thinks that the source of a female's power is obvious. She has something the male wants: an unfertilized egg. 'The female has this egg and she can say – you want to fertilize this? Guess what? If you want this, then I eat first,' Lewis told me.

Lemurs have an especially short breeding season – in white sifaka it's just thirty minutes to ninety-six hours a year, and in ring-tailed lemurs four to twenty-four hours. 'In terms of economics, if there's only one female in oestrus she should have a whole lot of power because short supply means high demand,' Lewis explained.

But evolution has other possibilities. As soon as you have just one or two females in heat at a time it incites males to compete to try to dominate this valuable resource. Males will evolve greater size and weaponry in order to fight off their love rivals, which, as a by-product, results in the kind of physical might that then allows them to dominate females, and subsequently reduce females' egg-resource clout. So the same forces that allow the female more dominion also select for sexual dimorphisms in the male, which then physically undermine female power.

This is Darwinian sexual selection 101, and red deer, *Cervus elaphus*, are a textbook illustrative case. The hinds all come into heat for a short period every year, inciting the males to rut. Stags have evolved overbearing antlers and increased bulk in order to compete with one another, which in turn makes them physically dominant over females.

In the case of the white sifaka, males do indeed compete physically for females, and battles can become quite bloody. But for some reason this has not resulted in them physically dominating females, as Darwin would have predicted. Lewis thinks this could be to do with Madagascar's peculiar environment and the sifaka's strange form of locomotion. A recent study revealed that when you live in the spindly dry forest, agility trumps brute force. If you're being chased by a competitor and you are too big, you're too slow and you'll get caught. And if you're too small, when you do get caught you won't be able to fight. So, selection favours an intermediate body size and powerful long legs, explaining why competitive males have not evolved greater size and the female lemur retains her power and social dominance.

There are further evolutionary forces at play. Just as we saw with the genitals of ducks, sexual conflict is also involved. Lemur females are promiscuous but males have evolved a sneaky trick to monopolize their valuable egg without physical dominance or fighting one another: they have semen that hardens like rubber to form a 'copulatory plug'. When females are only receptive for a short period, a male can temporarily enforce her chastity by clogging her vagina with his coagulated seminal fluids. These plugs can get pretty big – in ring-tailed lemurs they're over five cubic centimetres.* They don't make it impossible for subsequent males to mate, but they need to be dislodged, so they are a significant obstacle. When the female is only receptive for a day or even less, this might make all the difference.

A recent study by Amy Dunham, associate professor of ecology and evolutionary biology at Rice University, found that copulatory plugs are more common in species with short fertile windows (like the lemurs) and no difference in size between the sexes. She believes this alternative form of mate guarding offers an additional explanation for why male lemurs didn't evolve to physically dominate females with increased size and weaponry.

This monomorphism could, according to Dunham, be the key to understanding female dominance – the so-called 'Holy Grail' of lemur research. Game theory predicts that when two contestants are equally matched, the winner will be the one that values the prize most. Females, with their elevated reproductive costs, have higher nutritional demands than males and are more at risk from going hungry. Undernourished females are unlikely to produce quality eggs or support pregnancy and lactation, but a skinny male can still shoot viable sperm and fire his genes into the next generation. So, females have more to lose in the reproductive fitness stakes and will therefore be expected to fight harder for resources.

---

* Alan Dixson and Matthew Anderson of San Diego Zoo created a catalogue of primate sex plugs complete with an official scale of semen coagulation from 1 (no coagulation) to 4 (solid semen plugs). They found an interesting pattern: the more promiscuous the female the more solid the plug. For what it's worth, humans scored a 2: 'semen becomes gelatinous and remains semi-fluid but there is no distinct coagulum'.

Physical fighting is also costly and so it pays for males to defer to females and search out more food elsewhere rather than enter into an extended battle that they're unlikely to win and that could instead cause significant harm. Given that most lemurs are the same size, and food is so scarce on this harsh and highly seasonal island, this could explain why those male sifaka repeatedly give up their precious baobab fruits after a couple of cuffs to the head.

Female lemurs are also wired for aggressive competition. Christine Drea, the Duke professor we met in chapter one studying the spotted hyena, has noted that many lemurs also share the same giveaway physical quirk: 'masculinized' genitalia.

The spotted hyena (*Crocuta crocuta*) is a candidate for most domineering female mammal on the planet. They aggressively overpower males in most situations and sport an eight-inch clitoris that's shaped and positioned exactly like the male's penis. They also have a false scrotum and no external vaginal opening. Instead they must copulate and give birth through their 'pseudo-penis'.

Female lemurs are less extreme, but still sport some pretty funky junk. Sifakas and dwarf lemurs have a vagina that's only open for business for a day or so during their brief breeding season, sealing shut for the remainder of the year. Some lemur species have a pseudo-scrotum with 'skin identical in composition to that of the male's scrotum' and many boast a clitoris that superficially resembles a penis: elongated and pendulous yet stiffened by erectile tissue and an internal bone. The ring-tailed lemur's clitoris is as thick and almost as long as the penis, with a urethra that sits within the shaft, allowing females to urinate from the tip, like a male.

Ring-tailed females could 'write their name in the snow', joked Drea over Skype. This isn't just a neat party trick, but a dead giveaway for specific hormonal activity. 'It's very unusual,' Drea explained, 'and a hallmark trait of exposure to androgens.'

Sure enough, just like the spotted hyena, pregnant ring-tailed lemurs reveal elevated levels of testosterone along with the lesser known androgen androstenedione, or A4. Drea and her team recently discovered that in ring-tailed lemurs the levels of A4 during pregnancy can even predict the dominance of her resulting daughters. During a recent long-term study they measured the concentration of A4 in pregnant females and then later monitored the subsequent level of rough and tumble play experienced by their offspring. 'If mom's A4 is high, her daughter won't be the target of so much aggression,' Drea told me. 'Basically, she'll rule.'

It would seem that soaking in a prenatal androgen soup wires these fetal females towards aggression, giving them the competitive edge as adults. But if evolution is all about maximizing genetic posterity then this antagonistic advantage is something of a double-edged sword. For the spotted hyena, increased aggression may help her and her cubs fight off rivals when feeding is highly competitive at a communal carcass. But the price is giving birth through a clitoris, which is not without its challenges. For first-time mothers it's the eye-watering equivalent of squeezing a cantaloupe out of a hose pipe, which is why up to 60 per cent of births are stillborn and 10 per cent of new mums die in the process.

'There's a long string of negative consequences to females being exposed to androgens' – Drea explained that as adults andro-genized females can experience difficulties with reproduction and mothering – 'so what you have in these female-dominant species is that they have to figure out a way to reap the positive consequences of androgen exposure while minimizing the deleterious.'

It's a fine balance. The compromise of intensified androgen exposure is evident in Drea's female-dominant study subjects. Like the spotted hyena, ring-tailed females and their offspring may benefit from violently commandeering precious food sources in an other-wise barren Madagascan dry forest. But their aggression levels are so high that lemur mothers frequently end up brawling so hard with other females that their own infants wind up as casualties of war, which is less than ideal.

A wide range of female mammals, from meerkats to the common garden mole, sport some degree of 'masculinized' genitalia. Most interesting to Drea is its presence throughout the prosimians – the group of primates that includes Madagascar's lemurs, along with

Asia's lorises and Africa's bush babies. All these primates come from the most primitive primate line that diverged from New and Old World monkeys some seventy-four million years ago. This suggests to Drea that androgen-mediated female dominance could have been the ancestral state of not just lemurs but all primates, which includes us.

Dr Lewis has come to the same conclusion. In an as yet unpublished paper she has mapped out a family tree of living and extinct species' physiology and deduced that the common ancestor of all the lemurs and, in fact, all primates, was likely to be monomorphic. Male dominance is associated only with larger males, which suggests that our shared ancestor must have been either co-dominant or fully female dominant.

This revolutionary proposition destroys the assumption that aggressive patriarchy is the universal nature of all primates. It makes total sense to Drea, who thinks we've been looking at the 'puzzle' of those bossy females through the wrong lens.

'Everybody always says, "Why would you have female dominance?" Well, why wouldn't you?' she exclaimed over Skype, with seasoned exasperation. 'You have a placental mammal in which the cost of reproduction is borne by the female. Why would you not have a situation arise where that female would have advantages over the male?' The answer, she believes, is if dominance is achieved by increased aggression from androgen exposure, then this comes with some costly side effects that only some species have evolved to manage and thus maintain their reproductive fitness. Other females have discovered ways to wield power that thankfully don't involve giving birth through a clitoris.

SISTERS UNITE!

'People think power is all about fighting,' Rebecca Lewis explained to me one night in Madagascar as we dined on rice and dried fish at our campsite, 'but not all power comes from fighting.'

Power in animal societies has traditionally been defined in terms of dominance through physical intimidation – which is a very male way of looking at it. Lewis believes we need to find a new way of categorizing power structures to recognize the commanding influence of females that are small yet mighty.

'I grew up in Mississippi, in the South, where you'd never say that women are dominant,' she told me. 'But women have a whole heck of power in that you don't mess with somebody's mother, sister, wife or daughter. So I grew up understanding that there are different kinds of power.'

As Lewis sees it, power can come from physical dominance or what she calls 'economic leverage'. This could be specialist knowledge of where to find the best fruiting trees, controlling access to unfertilized eggs or strategic alliances.

The celebrated Dutch primatologist Frans de Waal, professor of primate behaviour at Emory University in Atlanta, Georgia, agrees that female power is underrated. I caught up with him while he was in London and he told me about the far-reaching influence of Mama, the alpha female of a colony of captive chimpanzees he studied in Arnhem, Netherlands.

'Mama was a king-maker,' de Waal told me.

In chimps, the alpha male is officially the dominant political figure. But no alpha male could rise up and dominate the colony without Mama's support, which gave her an enormous amount of power. The males may have sucked up attention with their screaming and fighting, but Mama was, without doubt, 'the boss'.

De Waal first met Mama early on in his career in the 1970s, when the budding primatologist found himself drafted to Arnhem to document the social hierarchies of the freshly established experimental captive chimp colony. He quickly noticed the central figure was a female chimp with the air of a grandmother who'd seen it all and didn't take any nonsense from anybody. 'There is great power in Mama's gaze,' he wrote in his first book, *Chimpanzee Politics*.

Mama commanded respect from every ape in her orbit, including de Waal. The first time he locked eyes with her face to face she made the lofty six-foot-plus primatologist 'feel small'. Mama was broadly built and could be physically intimidating but de Waal noted she also had a sense of humour. She connected easily with all chimps, be they male or female, and developed a support network like no other in the colony.

Mama was the top-ranking female of the troop, the alpha female – a position she held for over forty years, until her death. De Waal believes her status came from her unique charisma and social skills. In chimpanzees, female rank is governed by age and personality. When females are placed together in zoos, they generally decide rank quickly and with little fuss: one female will show her submission to another and that's it. Female hierarchies are stable and rarely contested. According to de Waal they're maintained by 'respect from below rather than intimidation and strength from above' and perhaps are better described as subordination hierarchies.*

Amongst male chimps it's a very different story. Rank is determined partly by physical strength but crucially by tactical coalitions with other males. Alpha male status is frequently challenged and highly unstable. Power struggles generally involve complex and shifting alliances that de Waal has compared to human political manoeuvring.

---

* This term was originally coined in the 1970s by the brilliant primatologist Thelma Rowell, who felt there was too much emphasis on dominance, and less interest paid to the key role of submission – a preoccupation she attributed to the 'unconscious anthropomorphism' of primates by her male colleagues. Like Alison Jolly, Rowell is another unsung female primatologist who challenged conventional thinking, sometimes in provocative ways. It is telling that Rowell chose to disguise her sex by signing her published papers inconspicuously as T. E. Rowell, but this simple disguise still caused issues. In 1961 she submitted a paper to the Zoological Society of London Journal. The society was impressed and invited T. E. Rowell to come down from Cambridge and give a talk to the fellows; but when it was discovered that T. E. Rowell was, in fact, a woman, there was some embarrassment. She was able to give the lecture but not to sit with the fellows for dinner because of her sex. The solution was to ask her to sit behind a curtain, out of sight, and eat her meal. Needless to say, she declined.

When tensions in the Arnhem colony reached boiling point the combatants always turned to Mama, who de Waal has described as 'a born diplomat'. She would have no fear in entering the fray of warring males, where she would purposefully groom the underdog and tensions would soon evaporate.

Mama's power came from her command of the sisterhood. Every male in the Arnhem colony knew they needed Mama on their side because she represented all the females. This made her a powerful ally, but she was far from impartial. She'd take sides in male power struggles, choosing to support one male against another. If one female in the group dared support the wrong male, they'd find themselves in trouble with the boss until they switched allegiance to Mama's favoured candidate.

'She was basically a political whip – keeping everyone in line,' de Waal explained to me.

The males all knew they had to keep Mama sweet. They would groom her and tickle her babies. If Mama snatched food from their hands, they never grabbed it back or complained. They put up with whatever she did, just to have her on their side.

'We need to move away from this idea of dominance as being something special,' de Waal told me. 'We need to make distinctions. You have physical dominance, which clearly in many species is male. Then you have rank, which is communicated more between males and between females than between the sexes.'

Rank is measured by who submits to whom, which chimps do by bowing and pant-grunting. These outward signs of status reflect what de Waal calls the 'formal hierarchy' and act like military stripes on a uniform.

'Finally, you have power,' de Waal told me, 'meaning how much influence you have on the social processes in the group, which is much harder to define.'

De Waal believes that power hides behind the formal order. Social outcomes in a chimp group depend on who is the most central in the network of family ties and alliances. With her superior social network and mediation skills, Mama was extraordinarily influential. All the adult males formally outranked her, yet they all needed and respected her. 'Her wish was the colony's wish,' says de Waal.

The king-maker power of alpha females has been observed in other classic 'male-dominated' primates. Barbara Smuts, the distinguished professor of anthropology at the University of Michigan, has documented how in rhesus macaques and vervet monkeys, for example, a male's quest to achieve and maintain dominance is strongly influenced by the support of high-ranking females.

Female vervet monkeys remain in their birth group and form strong lifelong bonds with their kin, while males disperse and join other, unrelated groups. This gives the females a huge amount of power. Matrilines of related females form the stable core and cooperate against male domination. Females will prevent certain males from joining the group and drive other males out, occasionally wounding or even killing them in the process.

'They are essentially female groups. Males may float in and out, but the alpha female is the central figure and has a lot of power,' de Waal told me.

As the stable nucleus of the group, females are often its brains; the keepers of essential environmental knowledge about where the best spots for sustenance or a safe sleep can be found. Female mammals often live longer than males, enhancing their expertise. Such wisdom provides females with the authority to lead their group. Amongst capuchin monkeys, for example, it is the diminutive females that more commonly display leadership when it comes to foraging and group movements not the alpha male, challenging the age-old assumption that dominance and leadership are one and the same.

The social influence of these female matrilines was overlooked for decades, with researchers focused instead on the more dramatic politics of the alpha male and the noisy shenanigans of his dominance hierarchy. Female primates were considered to be too

preoccupied with mothering to be capable of organizing themselves into any kind of power structure. 'Primate females seem
biologically unprogrammed to dominate political systems,' wrote
Lionel Tiger, the celebrated Canadian anthropologist famous for
coining the term 'male-bonding', in 1970.

These stereotypes are slowly shifting. The more female matrilines
are studied, the more group authority they command, influencing
societal outcomes in ways that have traditionally been underappreci-
ated when viewed through the prism of physical dominance, and
thus chipping away at the assumed autonomy of the alpha male.

What's interesting about Mama's story is the power she drew
from a social network consisting of *unrelated* females. The colony at Arnhem, although striving to be as natural as possible, was
artificial in that it brought stranger females into close proximity
with one another. It also provided them with ample food, so they
didn't have to compete for resources, giving the females a chance
to bond.

This experimental situation demonstrates how flexible chimpanzee social roles are and how easily they adapt to different
environmental situations. In the wild female chimps do not enjoy
such affiliative authority. Unlike female vervet monkeys, once a
female chimp reaches puberty, she leaves her birth troop and sur-
renders to a nomadic life, foraging alone in the forest. Any females
she meets along the way will be viewed as competitors, so they
don't bond. If she happens to join a group, there will be no familiar family members to connect with and her only significant
connection will be with her offspring.

In contrast, males do not disperse and spend their lives surrounded by their family. Male chimps form the core of the troop,
and over the course of their lives they develop complex relationships and supreme social muscle.

Dispersal patterns, then, can be a neat way of predicting the
dynamics of power in social primates. The influential Harvard
University anthropologist Richard Wrangham formulated this
observation into a much-cited theory that expects the sex that

stays in their birth group to always develop the strongest mutual bonds. Wrangham based his seminal thesis on the handful of primates where females, as opposed to males, migrate in this way. Sentenced to a pitiful, powerless Billy-no-mates lifestyle, non-bonded female primates are every feminist's nightmare. Wrangham notes they're all subjugated by physically superior males, suffer 'unimportant and undifferentiated' relationships, have little evidence of coalitionary alliances and no apparent dominance hierarchy of their own.

In addition to chimps, non-bonded female primates include gorillas, colobus monkeys and hamadryas baboons (*Papio hamadryas*) – whose females are the least emancipated of the bunch, with the dubious honour of being dubbed 'the most wretched and least independent of any non-human primate' by the anthropologist Sarah Blaffer Hrdy.

You do not want to be reincarnated as a hamadryas female. These social Old World monkeys live in large herds and scratch out a meagre life foraging for seeds and shoots in the semi-desert badlands of Somalia, Sudan and Ethiopia. Sexual dimorphism is extreme: burly males are twice the size of females, sport horrifying canines and a splendid white mane. Females, by contrast, are nervous-looking, scrappy little brown creatures.

Virile males amass and maintain a harem of around ten to twenty females in a uniquely creepy way, by kidnapping them from their families as pre-pubescent juveniles. From day one, the immature female is conditioned by her captor to unfaltering obedience by routine domestic violence. If, for example, she strays even a few metres for an impromptu sip of water, she will be hounded and attacked by her subjugator, occasionally with such force she's lifted clean off the ground. These 'excessively paternalistic' males rarely cause grave harm to their hostages, however. Their hostility is carefully tuned to intimidate and control, without inflicting irreparable damage to their precious reproductive investment.

There are very obvious parallels here with human societies. The communities in which women have least control over their lives, and are at most risk of male violence, are ones in which they are separated from their kin at an early age and have little support.

Wrangham's non-bonded females are a boon for anthropologists looking for evidence of a natural state of aggressive male ascendancy in some of our closest cousins. More depressing still is Wrangham's assertion that patrilocality – males stay, females leave – was probably the pattern of pre-humans, implying that our female ancestors would have been similarly isolated, vulnerable and oppressed.

There is, however, one primate that rips apart Wrangham's law and rescues hope for a more female-empowered vision of our past and our future: the bonobo.

Frans de Waal has called the bonobo 'a gift to the feminist movement'. Where chimps are patriarchal and warlike, bonobos are matriarchal and peaceful. We are equally related to both. The unorthodox lives of these obscure great apes provide the final death blow to the notion that aggressive male dominance is cemented into primate life.

Bonobos are the rarest of the five species of great ape. They're only found in the lush rainforests on the southside of the Congo River in the Democratic Republic of the Congo and their numbers are small – less than fifty thousand individuals in a territory of less than 500,000 km^2.

The remoteness and political instability of the bonobos' homeland combined with their small population size ensured their anonymity well into the twentieth century. They are, in fact, one of the last large mammals to be described by science. The bonobo was first discovered by taxonomists in the dusty archives of a Belgian colonial museum. It was 1929 and a German anatomist by the name of Ernst Schwarz was examining a skull that had, on account of its diminutive size, been described as a juvenile chimpanzee. There were, however, small yet unmistakable differences that niggled the anatomist. When he realized the skull belonged to an adult, Schwarz announced to the world that he had stumbled upon a new subspecies of chimpanzee.

Originally classified as the pygmy chimpanzee, *Pan paniscus*, bonobos are now recognized as their own distinct species. They do indeed look a lot like their chimp cousins – if a little smaller, more lissom and with less hair. Like chimps, females are around two thirds the size of males and also migrate from their birth group. Their social lives could not be more different, however. Instead of living out their adulthood as a forlorn diaspora with little or no agency, females join groups and form alliances with unrelated females. The power of this constructed sisterhood allows them to dominate the bigger males. It is formed and maintained not by fighting and physical intimidation, but by what scientists describe as G-G rubbing, shorthand for genito-genital rubbing. In other words, female bonobos have evolved to overthrow the patriarchy by perfecting the art of mutual frottage.

On arrival in another community, a young bonobo female will single out one or two senior resident females for special attention, using frequent G-G rubbing and grooming to establish a relation. If the residents reciprocate, close associations are set up, and the younger female gradually becomes accepted into the group – then, after finding a mate and producing her first offspring, the young female's position becomes more stable and central.

This form of sex is undocumented in other wild primates. But female bonobos indulge in G-G more frequently than any other kind of sexual activity. It's their go-to social lubricant and serves to increase social standing, promote cooperation and regulate any competitive tension amongst unrelated females, especially during feeding situations. They certainly look like they enjoy it. Females engaging in G-G emit the kind of grins and squeals that indicate they're having a most excellent time indeed.

'I'm pretty sure they orgasm,' Amy Parish told me. 'Their clitoris is actually positioned to get maximum stimulation from this kind of sex.'

Dr Parish is the scientist who first exposed the bonobo female's extraordinary solidarity secret. I met her on a hot summer's day at San Diego Zoo, which is home to the group of bonobos that helped lead her to this revelation. The self-confessed feminist Darwinian has an extraordinary CV, having been mentored by a roll call of the greatest living primatologists. She's fearsomely smart, but won me over immediately by arriving at our meeting rocking a pair of pink sunglasses shaped as love-hearts.

Parish chose to study bonobos for her PhD and has been documenting their social lives for thirty years. Back then precious little was known about this petite great ape. Parish set herself up observing their behaviour at San Diego Wild Animal Park, where she quickly noticed the unique nature of female friendships, which contravened Wrangham's law.

'What was fascinating to me was how close the females were to each other: hanging out, playing and being nice to each other's infants,' she told me. 'We expect females to get along when they have the benefit of kinship, but we don't see that in mammals when the females are unrelated. You see them either being avoidant, antagonistic or downright aggressive.'

Feeding time can be especially fraught. But not with bonobos. As we sat on our viewing platform in San Diego Zoo observing supper time, Parish pointed out how Loretta, the alpha female of this group, was in control of the food. Access is often provided in exchange for sex. Both males and females will trade sex for food and, as a result, happily sit and feed together. This is in sharp contrast to chimps, in which males eat first with females sitting at a safe distance until they are fully sated.

'I think that sex helps to lay a foundation that dispels any tension, which would get in the way of forming longer-term bonds,' she told me.

Unrelated females do indeed form long-term stable relationships with one another, facilitated by grooming and sex. Parish noted they were also backing each other up and forming coalitions. Unlike male chimps, they don't use their coalitions to fight one another but to overpower aggressive males.

Parish noticed that females were inflicting serious blood-drawing injuries on males – deep gashes, fingers and toes bitten off and she even witnessed punctured testicles. Frans de Waal, her supervisor, sent her a list of twenty-five injury cases that he had recorded whilst studying the bonobos in San Diego. Almost all were female-on-male attacks. Parish cast the net wider and discovered shocking tales from zoos all around the world. At Wilhelma, a zoo in Stuttgart, Germany, for example, two females attacked a male and bit his penis in half (a microsurgeon repaired the damage and the male went on to reproduce).

Each zoo had their own folkloric story about what was 'wrong' with their males because this pattern of aggression wasn't what people thought was 'natural'. Parish looked at the data through a different lens and came to the landmark realization: this was a species with female dominance. 'That had never been described for bonobos before.'

I saw a glimpse of this in the San Diego group. Lisa, a female with leadership aspirations, had been demonstrating her authority by bullying Makasi, a low-ranking male. Fingers had been bitten and blood drawn on a number of occasions, so Lisa and Makasi were now in separate groups to prevent further injury.

'It's really clear that the females are not messing around. It's serious and it's dangerous for the males and they have a lot of fear of females,' Parish told me.

The result is that male bonobos are much less aggressive than chimpanzees. 'They probably have learned their lessons over the years,' de Waal told me.

Male bonobos are very close to their mothers, whose rank and authority offer their sons protection against bullying by other females. Makasi's mother was not at San Diego Zoo, leaving him vulnerable to attack. In the wild, males will likely have their mother close by in their social group. So, whilst the threat of female aggression is very real, bonobos are actually much more peaceable than their chimpanzee cousins.

Chimpanzees are famously territorial. When neighbouring groups meet, the scene is extremely hostile: males tear about with their hair on end, their body language set to intimidate. They scream, bang on trees and will even kill each other. In stark contrast, when bonobo groups meet there is no sign of fighting.

'They might shout a little bit in the beginning, but very soon it looks more like a picnic than warfare,' Frans de Waal told me. Albeit a picnic where everyone is having sex with each other.

'Sex is the bonobo's answer to avoiding conflict,' de Waal added, which is why these unconventional apes have been dubbed the 'make-love-not-war' hippy ape.

The bonobo's sex life is as creative as it is unrestrained. Males, for example, will enjoy each other's company by 'penis-fencing', which sees them rubbing their 'swords' together whilst dangling from a branch (nice work if you can do it). For females, the most frequent and preferred sexual activity is G-G, which they will choose over sex with a male if both present themselves at the same time.

'There's no bonobo that's exclusively heterosexual or homosexual. They're all bisexual,' Parish told me.

Like humans, bonobos have a partial separation between sex and reproduction, with females frequently initiating and engaging in sex outside of their fertile period. But with the average copulation lasting thirteen seconds, bonobo sex is fast, frequent and seemingly as casual as a handshake is to us.

It's also remarkably familiar. Bonobos stare deeply into each other's eyes, passionately kiss using tongues, practise oral sex and even fashion sex toys. Frances White, a biological anthropologist at the University of Oregon, once watched a female bonobo turn a stick into a kind of knobbly French tickler, which she then proceeded to take great pleasure in.

But it is the most conservative end of their sexual repertoire that caused the biggest stir. When bonobos do have heterosexual sex, they often do so in the missionary position. This is not really seen in any other primates. Chimpanzees virtually never have sex face to face, whereas bonobos do so in one out of three copulations in the wild.

The first suggestion that the sexual behaviour of bonobos in some ways mirrored our own arrived way back in the 1950s, but the scientists involved chose to obscure their controversial findings by reporting in Latin. Eduard Tratz and Heinz Heck reported in 1954 that the chimpanzees at Hellabrunn Zoo mated *more canum* (like dogs) and bonobos *more hominum* (like people). In those days, face-to-face copulation was considered uniquely human, a cultural innovation that needed to be taught to preliterate people (hence the term 'missionary position'). These early studies were studiously ignored by the international scientific establishment. It wasn't until the sexual liberation of the 1970s that the full glory of the bonobos' sex life started to go public.

The bonobos' novel approach to harmony and hierarchy has been observed in the wild, as well as in the more artificial social situations afforded by zoos. Female bonobos in Congo's LuiKotale forest have even been recorded using specialized gestures and pantomime to convey their desire for a bit of G-G. The soliciting female will point backwards with a foot towards her sexual swelling and then shimmy her hips in imitation of a rub, at which display the second bonobo will embrace her for the real thing. The authors of the paper note the significance of such gestures in the evolution of language – the ability to point has also been linked to facilitating cooperation and behavioural coordination in humans.

Like chimpanzees, bonobos share almost 99 per cent of their genetic make-up with us. They both have equal right to claim status as our closest cousin. The ancestor of chimps and bonobos diverged from our ancestral line a mere eight million years ago. The two *Pan* species then diverged from one another much later, which is why they appear to be so much more similar to each other than to us.

Frans de Waal has suggested that, if this evolutionary scenario of ecological continuity is true, the bonobo may have undergone less transformation than either humans or chimpanzees. It could most closely resemble the common ancestor of all three modern species. Indeed, in the 1930s Harold J. Coolidge – the American anatomist who gave the bonobo its eventual taxonomic status – suggested that the animal might be most similar to the primogenitor – our shared ancestor – as chimpanzee anatomy shows more evidence of specialism through evolution.

Bonobo body proportions have been compared with those of the Australopithecines, a form of pre-human. As I watched the bonobos at San Diego Zoo I was struck by how, when they stand or walk upright, they look as if they have stepped straight out of an artist's impression of early hominids – especially the matriarch Loretta, who exuded a sage, and remarkably naked, authority.

Compared to the alpha male primates living in nearby enclosures – the enormous orangutan, with his long swinging ginger dreadlocks and huge facial flanges, or the terrifying musculature of the silverback gorilla – Loretta certainly wasn't physically impressive. In fact, in the words of Dr Parish, 'she looks a bit like Shrek' on account of her sticky-out ears and being virtually bald. What appeared to me to be a tragic case of alopecia was actually a sign of the matriarch's high status. Being groomed results in hair loss, so the higher ranking you are, the less hair you have. The resulting near-naked skin made Loretta look so much more human than any hirsute chimpanzee ever could.

It was Loretta's reaction to Dr Parish that really bridged our genetic gap and sent shivers down my spine. After the scrum of feeding time, the bonobos settled down and began to notice the humans gawping at them from the other side of the glass screen. The bonobo keeper commented how she always feels special if Loretta gives her a nod and I wondered how the ageing matriarch would register Dr Parish's presence.

The two females first met in 1989, when Parish was a fledgling PhD student and Loretta a youthful matriarch. Parish went on to spend all daylight hours, seven days a week, for several years documenting Loretta and her group. Since that study she'd paid regular visits. The primate and the primatologist had watched each other mature, become young mothers and practised elders. So I was anticipating some form of recognition, but I was blown away by what actually happened.

Loretta clocked Dr Parish and made an instant beeline towards her. The bonobo stood upright on the other side of the glass, her soulful amber eyes stared deeply into Parish's and she made a succession of subtle nods with her head. Parish nodded back using the same shared language of acknowledgement. Loretta then leaned into the glass, placing her head against it. Parish did the same and the two mock-groomed one another through the glass for over twenty minutes. At one stage Loretta placed her hand on the glass and the scientist placed hers against the bonobo's, as if the glass were not there.

It was a profoundly moving scene. I felt a lump in my throat. I was not alone. The surrounding zoo visitors were equally transfixed by these two old friends showing their love for one another. We were all awestruck into a respectful silence, and, in my case, trembling. Afterwards Parish told me I had indeed witnessed something special. She and Loretta had not seen each other for over a month; they didn't normally have such intensely emotional and protracted greetings.

I marvelled at how privileged Parish was to have experienced this connection and to have shared so much history with an animal that is so very close to being human but isn't. It was an intensely special relationship. This wise old female had helped Parish decode the secrets of her peaceful matriarchal society, and helped us humans understand, once and for all, that patriarchy and violence aren't necessarily burnt into our DNA.

This radically different, equally closest living relative forces us to rethink models of human ancestry to accommodate real and meaningful relationships between unrelated females, flexible social systems in which dispersal patterns do not necessarily dictate the bonding potential of females, and the possibility of systematic female authority over males, even if they are the *physically* dominant sex.

The bonobos opened anthropology up to explore fresh models that don't assume patriarchy as an ancestral universal state. In fact, its apparent rarity amongst the full cast of our primate cousins poses the more interesting question of how and why patriarchy evolved and took grip in many human societies.

Parish's former supervisor, the brilliant Barbara Smuts, incorporated bonobos into a new thesis which proposed how, over the course of human evolution, this unusual degree of gender inequality came about. She pointed the finger at our ancestors' gradual switch from hunter-gathering to intensive agriculture and animal husbandry. While cooperation on hunts gave men the possibility of controlling food resources, women's contribution to foraging limited this control. The smaller plots of land associated with the switch to intensive agriculture and animal husbandry, however, limited women's movement and gave males control over resources and an incentive to form political alliances with other males, to fight rival males and control females.

A foraging lifestyle makes it much harder for males to restrict females' movements and access to resources, as they are able to source their own. Once females were restricted in their activities and males gained control of high-quality foodstuffs, like meat, females lost agency and became sexual property. Paternity became an issue, as property was inherited, and patriarchy took hold. The evolution of the capacity for language allowed males to consolidate and increase their control over females because it enabled the creation and propagation of ideologies of male dominance/female subordinance and male supremacy/female inferiority.

'The roots of patriarchy lie in our pre-human past,' Smuts says. 'But many of the forms it takes reflect uniquely human behaviours.'

Not all anthropologists were as willing to embrace the bonobo and reconsider human history so readily. 'Some of my chimpanzee-studying colleagues were not exactly thrilled,' Parish told me. 'For forty years they'd cornered the market on "man's closest living relative". All our models of human evolution were based on physically aggressive, male-bonded and male-dominated chimps.'

Academia is a hotbed of competition and ego, with researchers jockeying for their animal and research to be the most relevant. If you've built a career around proving that humanity's patriarchal roots pass through chimpanzee culture and beyond, it's not easy to write off a lifetime of data and start afresh.

'I think it came as a shock, as it was so contrary to what we think is "natural" in the world,' Parish told me. 'There was a lot of sexism in people's reactions. Some of my male colleagues didn't want to recognize bonobo society as female dominant.'

Frans de Waal agrees with Parish – the primatologists who want to marginalize the bonobos are never women. 'It's all men,' he told me, and then illustrated this statement with a funny story about the infuriated reaction of a distinguished male biologist to one of de Waal's bonobo lectures.

'There was an old German professor who stood up and said' – de Waal adopts an incensed tone – ' "What is *wrong* with those males?" I explained there's nothing wrong with them. They have a good life. They have a lot of sex and I don't see what's wrong with them. But he was *really* worried about them.'

In the same way that Alison Jolly's revelations about lemurs were swept aside, bonobo female dominance was similarly diminished and redefined by many primatologists as 'male chivalry' or 'female feeding priority coupled with male social dominance'.

Craig Stanford, a celebrated chimpanzee careerist at the University of Southern California, was particularly vocal. 'He argued that it's not female dominance – it's strategic male *deference*, which the males do to get more sex,' Parish told me. 'He really wouldn't give up on that idea and it's still in his textbook. That was pretty irritating because it was just so outrageous.'

Some have even gone so far as to dismiss any female influence over the traditional patriarchal models of hominid evolution as a 'politically driven illusion engendered by feminism'.

Today only a few detractors persist. Most agree that in captivity female bonobos are always dominant over males. In the wild, de Waal explained, the hierarchy is more mixed, but the top spot is usually a female or two, then perhaps a male. Most males are subordinate to most females.

'Just imagine that we had never heard of chimpanzees or baboons and had known bonobos first,' Frans de Waal has wryly noted. 'We would most likely believe that early hominids lived in female-centred societies, in which sex served important social functions and in which warfare was rare or absent.'

In the end, the most successful reconstruction of our past is probably a mix of chimpanzee and bonobo characteristics. Whether it's more bonobo than chimp can be argued for ever, and probably will be. But to me that's not what matters most. The past is the past and cannot be changed. The future, however, can. And that's why I find the bonobos so inspiring. Their story shows us that males are not genetically programmed to aggressively dominate females. Their ability to do so depends on environmental and social factors. The key ingredient for female empowerment is the strength of the sisterhood, from family to friends, to overthrow an oppressive patriarchy and foster a more egalitarian society.

Parish agrees. 'We have a lot to learn from bonobo females. The feminist movement argues that if you behave with unrelated females as if they are your sisters, you can gain power. The bonobos show us that that's true. It gives us a lot of hope.'

Amen to that.

CHAPTER NINE

Matriarchs and menopause: kinship with a killer whale

The futuristic skyline of downtown Seattle provided an unlikely backdrop for my first encounter with one of the planet's most powerful predators. But there it was – a geyser of white mist followed by an unmistakeable long, slender black dorsal fin, maybe six feet tall, slicing its way through the silvery waters of Puget Sound, the watery backyard of America's Emerald City.

The killer whales were in town and it genuinely felt as if I was in the presence of rock stars. The port of Seattle is the third busiest industrial harbour in the US, zigzagged by a cacophony of car ferries and honking monster cargo ships, but the whales cruised through rush hour with the kind of insouciance only available to a six-tonne killer.

The soundtrack from *Reservoir Dogs* looped through my head as around twenty-five of them, including a young calf, navigated the busy bay. When a gigantic freighter passed particularly close by they didn't disappear to the depths; instead, they took the opportunity to demonstrate their extraordinary charisma by surfing its bow wave.

It was quite the show. The orcas jumped and pirouetted out of the water having, quite literally, a whale of a time. I had goosebumps. I wasn't alone: the deck of the boat was crammed with wide-eyed whale-watchers waving cameras and gasping with every breach – an audible register of delight that's known in the whale tourism trade as an orca-gasm.

'You've hit the lottery,' Ariel Yseth, a local veteran orca observer told me. 'The southern residents haven't been seen together like this for months.'

I was privy to a rather special event: an orca party. Killer whales, *Orcinus orca*, are the most pumped-up member of the dolphin family and, like their smaller squeaky cousins, they're highly social creatures with the smarts to match. Their whopping seven-kilo brain has more surface area for complex thought processes like language, social cognition and sensory perception than any other animal on the planet.

They live in extended family groups or pods of around five to thirty individuals, and when familiar pods meet they're known to take part in 'a greeting ceremony'. Each group lines up in opposing rows, side by side, and hovers at the surface of the water, pod facing pod, for several curious minutes before erupting into a mosh pit of play.

The southern residents are a clan of three such pods – known as J, K and L – with a reputation for being especially frolicsome. The exuberance I was witnessing – cartwheeling and breaching – marked the first time the whole of J-pod and members of K-pod had been seen together in eleven months.

A TV helicopter hovered overhead. 'The whales are big news these days,' Ariel told me.

In the last few years the southern resident population have been listed as endangered. A precipitous decline in wild salmon stocks – their sole prey – is principally to blame. However, an increase in pollution, be it toxins that get stored in the orca's blubber or noisy maritime traffic that messes with their ability to echolocate their dwindling quarry, are compounding factors.

These so-called 'residents' (as opposed to the other more transient pods of orca seen in these waters) used to be a guaranteed daily sighting in the Salish Sea during summer months, but their movements are becoming increasingly unpredictable.

'Everything's changed since Granny died,' Ariel told me.

Granny, or J-2 as she was officially known, was the elderly 'grandmother' to the J-pod. This old-lady orca was also the leader of the southern residents, and Granny's command of this clan of seventy-odd orcas was evident even to human landlubbers. When she wanted the group to follow her lead and change direction, she'd rise up, slap her two-metre-wide tail fluke on the surface of the water and they'd all come swimming. 'It was her way of saying, "Come on, kids!"' Ariel told me.

When Granny disappeared in October 2016 estimations put her at somewhere between seventy-five and a hundred and five years old, making her the oldest killer whale on record. But the most extraordinary thing about this elderly matriarch wasn't her age. It was the fact that she stopped having babies around the age of forty, yet lived on for several more decades, enjoying a post-reproductive career that was at least as long, if not longer, than her reproductive one.

Menopause is extremely rare in the animal kingdom. Theoretically, it shouldn't exist at all. Natural selection takes a pretty merciless view of a loss of fertility. And with good reason. If the purpose of survival is reproduction, then there is no reason for an animal to stay alive when it can no longer parcel its genes in a neat fleshy package for the next generation. Famously long-lived animals such as Galapagos tortoises, macaws and African elephants all continue to reproduce well into their twilight years.

As such, we humans have long been considered menopausal freaks. Until recently, the only mammals we knew of to live beyond their fertility were in captivity. True menopause occurs when reproductive senescence becomes uncoupled from somatic senescence; meaning your sex organs age faster than the rest of your body. In the case of menopausal zoo dwellers, like gorillas, their lives have been artificially extended by free meals and healthcare. In the wild a female gorilla lives for around thirty-five to forty years, whereas in captivity they can even top sixty years. So their bodies and brains have outlived their ovaries. Of the five thousand species of mammal, the only ones we now know to go through menopause naturally in the wild are the four species of toothed whale – and human beings.*

It's an odd thing to find kinship with a killer whale.

As a woman of a certain age, wrestling with my own dwindling fecundity and its associated threat of purposelessness and invisibility, Granny's story of post-menopausal power spoke to me too loudly to be ignored. I felt an urgent need to meet a menopausal killer whale and discover what gave us this strange synchronicity.

On the surface of things we have little in common – we haven't shared a common ancestor for some ninety-five million years (a tiny shrew-like creature that gave rise to a wide range of mammals including whales, humans, bats and horses). How had this orca grandmother apparently defied natural selection *and* become the leader of the rock-star southern residents? What can Granny teach us about leadership and menopause in humans? And, perhaps more urgently, as we stare down the barrel of an eco-apocalypse, what does it mean to lose our matriarchs?

We owe most of what we understand about the lives of killer whales to the southern residents. They have been studied continuously for over forty years, but it took a while for the men researching them in those early years to recognize that their society was matriarchal. As Ken Balcolm, founder of the Center for Whale Research (CWR) and one of the original southern resident scientists, told me: 'The females were *meant* to be the harem.'

When studies on the southern residents began in the 1970s, researchers often found groups of orcas that were made up of a few adult bulls, who could be clearly identified by their impressive size (males can reach nine metres in length) and towering dorsal fins. These males would generally be seen escorting a handful of more

---

* The four species of toothed whale to go through menopause are: orcas, short-finned pilot whales (*Globicephala macrorhynchus*), narwhal (*Monodon monoceros*) and beluga whales (*Delphinapterus leucas*).

diminutive individuals – one to two metres smaller – with comparatively squat dorsals that were assumed to be females.

According to Howard Garrett of Orca Network – who also worked at CWR in the early eighties – it was generally supposed, based on studies of other herding marine mammals such as sea lions, that the males would have used their size advantage to aggressively assemble a harem of females, and that sooner or later someone would see the big bulls battle each other for dominance and/or bully the females into sex.

A few years of close observation went by but such antagonistic behaviour never materialized. Instead, something quite unexpected occurred. Some of the whales that were assumed to be females sprouted tall dorsal fins and 'apparently transformed', quite obviously, into males.* More surprisingly still, they didn't then leave the pod, but instead remained swimming alongside the other males and females.

'The realization slowly dawned that many of the "females" were actually juvenile males, and that even after they became adults they stayed close beside their mothers,' says Garrett. At first, he claims, there was a reluctance to accept this observation of their social set-up. No other mammal known to science maintains lifetime contact between mothers and offspring of both sexes. There is always a bias towards dispersal of one sex, and typically, in social mammals, it's the sons that leave. Some of these orca matrilines contained up to four generations of male and female orcas. Could this unique structure have something to do with the female's exceptional post-reproductive lifespan?

Darren Croft, professor of animal behaviour at the University of

---

* Orcas have a libertine approach to sex. 'They're basically all having sex with each other,' Dr Giles of Wild Orca told me. It is not unusual for pubescent sons to have their first sexual experience with their mothers or elderly matriarchs. And males are known to engage in 'sword fighting' with other males. Whether this is how they learn to wield their six-foot organs and master tricky, fluid four-dimensional sex or whether it's something they simply derive pleasure from isn't clear. But what is clear is that they are most certainly not females.

Exeter, thinks it does. Croft has been gripped by the riddle of menopause, and the social systems that underpin it, for the last decade. 'Clearly, from an evolutionary point of view, menopause is maladaptive,' he told me. 'So its existence fascinates me.'

The human menopause puzzle has generated dozens of theories and decades of debate. A popular explanation says that a post-menopausal woman, like those zoo-dwelling gorillas, has simply outlived her ovaries thanks to modern medicine; implying that menopause isn't really natural and women should bow out gracefully at around fifty along with our fertility.

Thankfully, the existence of menopause in hunter-gatherer societies puts paid to that theory. 'There is overwhelming evidence that menopause isn't an artefact of our increased longevity but is instead deeply rooted in our evolutionary past,' Croft told me.

Evolutionary explanations for menopause cover a vast spectrum of thought, and at one end we have what I like to call the 'Hugh Hefner Hypothesis'. This wasn't actually developed by the late loungewear-sporting Lothario, but I'm sure the old bunny-lover would have approved.

It states that female menopause is the evolutionary upshot of the human male's preference for younger females. The trio of male scientists from McMaster University in Ontario that proposed this profoundly dispiriting theory back in 2013 supported it with some snazzy mathematical modelling demonstrating how the male penchant for young skirt results in the build-up of deleterious mutations, causing older females' ovaries (if not their hopes and dreams) to shrivel up and die before the rest of them.

At the other end of the spectrum is the more enduring, and significantly more feminist-friendly, 'grandmother hypothesis'. Proposed in 1998, it posits that females who step out of the reproductive rat race mid-life and focus their energies on supporting their children (and grandchildren), instead of squeezing out yet more babies, significantly increase their offspring's chances of survival and, in turn, their own genetic legacy.

The anthropologist behind this theory, Kristen Hawkes, based her theory on observations of actual living hunter-gatherer societies rather than abstract mathematical models. She noticed that mothers from the Hadza of Tanzania faced a trade-off between the back-breaking job of foraging for starchy tubers and caring for new infants. But if grandmothers helped with the digging up and the sharing of tubers and berries, they were rewarded with healthier grandchildren who weaned at a younger age.

The detailed, decades-long documentation of the southern residents' lives and family ties, accumulated by the Center for Whale Research, provided Croft with an alternative menopausal model animal and, most crucially, the data set he needed to test out various theories.

When I asked him about the Hugh Hefner Hypothesis, he told me there was no evidence of orca males favouring sex with young females: 'I can't see any scenario where that would be adaptive for a male orca.' In fact, I've been told it is quite the reverse: post-menopausal female orcas have a decidedly cougar-like sex life and are often seen soliciting eager young pubescent males for sex.

Trawling through more than forty years of underwater footage, field notes and photo IDs of dorsal fins, Croft and his team did discover that the post-menopausal females were more often than not the ones swimming at the front of the pod, guiding their family to the best foraging grounds, especially when food was in short supply.

Aside from humans, killer whales are the most widely distributed predators on the planet. Highly specialized hunting skills have enabled these cosmopolitan killers to exploit particular prey types from the Arctic to the Antarctic. For example, orcas off the coast of New Zealand specialize in digging up and devouring stingrays. In Argentina, they surf on to shore to snatch sea lion pups from the beach. Along Alaska's Unimak Pass, they gather in May to ambush young grey whales, and in Antarctica they use synchronized swimming to generate waves that wash seals from the safety of the ice floe – and into their mouths. These particular orca races are known to ecologists as eco-species, as they're the same species but inhabit a specific geography and do not interbreed. More than that, they are known to 'speak' their own dialects and their specialist hunting techniques, passed down from generation to generation, have been compared to culture.

The southern residents hunt Pacific salmon, ideally Chinook (also known as king salmon), and an adult killer whale must chomp through 20–30 per day to stay healthy. The Salish Sea (which straddles the American–Canadian western border) is a traditional feeding ground where whales gorge on a feast of these big, fatty fish. The salmon are drawn there in large numbers before they swim up the tributary rivers across the Pacific Northwest to spawn.

Locating these ephemeral salmon hotspots requires a wise and wily hunter since they change with year, season and even tide. Orcas must decide whether to spend energy following the fish up river or hang out at the salmon deep-sea diner hoping for fresh stock. It's a complex cognitive job which has become much harder now salmon have to navigate an obstacle course of gigantic concrete hydropower dams on their journey to their spawning grounds. This, combined with warming waters and decades of overfishing, has seen salmon populations crash. When the fish are in short supply, only killer whales with years of experience know how to find them – and those are the oldest matriarchs.

'It's almost like in a city, where the takeaway shops are only open one night of the month and you need to know which takeaway is going to be open on which night of the month,' Darren Croft explained to me over Skype.

Orcas in captivity have been shown to have phenomenal photographic memories: remembering testing patterns after twenty-five years. Not only are these wise old-lady whales a living library of ecological and cultural knowledge, they're also incredibly benevolent: 'You'll see a sixty-year-old female catching a salmon, breaking it in two and then giving half to her thirty-year-old son. It's incredible,' Croft told me.

Despite the tough name, male killer whales are, according to whale experts, 'massive mummy's boys'. They spend most of their lives swimming a few feet from mum's side but, significantly, her hunting handouts help keep them alive.

Croft's team discovered that if a male orca's mother died *before* his thirtieth birthday, he was three times more likely to die the next year. If she passed away *after* he turned thirty, he was *eight* times more likely to snuff it during the following year. But if mum had gone through menopause, his odds of dying the next year went up by *fourteen* times. The data was irrefutable: sons whose mothers live a long time after giving birth to them have a survival advantage over those whose mothers die earlier, and that only becomes more true as both mother and son age. These post-menopausal orcas thus support the grandmother hypothesis posited by Hawkes.

According to Croft, there's still a problem with this theory: it doesn't explain why females would stop reproducing halfway through life. 'Look at elephants,' Croft exclaimed. 'The older females act as repositories for ecological and social knowledge, but they don't have a menopause.'

Elephant matriarchs are some of the most formidable females on the planet. They are the chiefs of their family group, carrying the wisdom to outwit lions, form political alliances with other female elephants, and remember ancient water sources during times of drought. These charismatic giants have much in common with killer whales (and indeed humans) – long lives, big brains, complex communication skills and a large, fluid social network.

Karen McComb, professor of animal behaviour and cognition at the University of Sussex, found an ingenious way of measuring the social knowledge and decision-making skills of these grand old dames. She headed to the Amboseli Elephant Research Project in Kenya, where the elephants have been monitored since 1972 – longer than any other elephant population and almost as long as the southern residents. She drove around with a loudspeaker and a tape deck, playing elephant families audio recordings of other elephants and then watching their reaction.

Nervous elephants stop what they're doing, bunch together in a defensive huddle and sniff the air for further olfactory intelligence on the intruder. Families with older matriarchs were much better at assessing risk and only did this disruptive behaviour when the calls were from unfamiliar elephants. Those families with young matriarchs were quick to react defensively and were, according to McComb, 'all over the place'.

Not only were the older matriarchs more likely to tell friend from foe, they could also distinguish between the roars of a male and a female lion. This is a vital skill because although these sound almost identical, their threat is anything but. Whereas female lions are normally the primary hunters, when it comes to snatching elephant calves only males, who are 50 per cent bigger, can afford to take on such sizeable prey.

The older matriarch's superior powers of discrimination keep her clan safe, relaxed and focused on that old priority: eating. McComb's research shows us that the quick-thinking and confident leadership of older matriarchs results in an increase in offspring, supporting the grandmother hypothesis.

So, you'd think that these old ladies would step off the baby-making treadmill and concentrate their energies on their existing reproductive investment. Especially when you consider how elephant pregnancies, at twenty-two months, are particularly physically gruelling and that their calves take five to six years before they're fully weaned. But, as incredible as it seems, matriarchs in Amboseli have been observed giving birth in their sixties. Their reproductive rate does slow down, but it doesn't suffer the same steep decline as we see in orcas and humans. Studies have shown that a female elephant's ovaries are in fact still functioning in their seventies, so in theory they could keep giving birth right up to their death.

In order for the grandmother hypothesis to work, there needs to be a hefty cost to reproduction late in life otherwise there's no reason for orcas, or indeed any animal, to stop. Croft and his collaborators have proposed that the key to the orca's menopausal mystery lies in their peculiar social set-up, and in conflict rather than cooperation.

Having kids is expensive, but with orcas there is an interesting disparity between the cost of sons and daughters. When a young female orca starts reproducing, around the age of fifteen, she needs 40 per cent more salmon in order to produce the rich milk required to suckle her calf. Therefore, when a 'daughter' reaches maturity she suddenly places a significant toll on the nutritional demands of the pod.

Sons are a different story. When pods mingle they will mate with females outside their matriline, and although they may remain surprisingly close to their mother even when having sex, their offspring will be brought up by the mother in that different matrilineal pod. So for a mother, feeding her son's genetic legacy comes at the expense of another matriarchal group, which is far cheaper than if she has a daughter. Evolutionary theory predicts therefore that mothers will indulge their sons far more than their mature daughters, and this was indeed demonstrated by a twelve-year study on food sharing in killer whales. Conversely, it also predicts conflict between mothers and their female progeny once the latter reach sexual maturity, as a result of some unique kinship dynamics.

When a female orca is born, her father is in another pod and so her relatedness to the males in her own pod is low. As she ages, her relative relatedness to the pod increases as she produces sons and grandsons. Mothers are therefore always more related to their pod than their daughters and granddaughters are. This asymmetry of relatedness promotes conflict between female generations breeding at the same time. Natural selection will favour young mothers, who have less stake in the success of the wider group, who aggressively compete for limited resources. This prediction was borne out by the observation that when orca mothers and daughters bred simultaneously, calves born to older mothers were almost twice as likely to die in the first fifteen years of life as those born to the younger mothers.

This *social* cost to late motherhood provides the evolutionary impetus for a female orca to stop breeding mid-life so she can invest in her sons and grandsons and stop competing with her daughters and granddaughters. That incentive doesn't exist in elephants because their sons, like most social mammals, eventually leave their birth group. So elephant females become *less* related to their group-mates over time or, at least, no *more* related. An elephant matriarch's best bet, then, is to carry on reproducing until she dies.

Croft thinks this 'reproductive conflict hypothesis' could provide the missing piece of evolutionary motivation for the grandmother hypothesis to work in other menopausal creatures, like us.

As we discovered in the last chapter, in ancestral humans it's thought daughters would have dispersed to join new families. Initially the young female interloper would have no relation to the group. But once she started to have children she would become increasingly related. As she got older, helping her daughter and granddaughters raise their children would become more genetically beneficial to her, especially since having more children would put her new kids in direct competition for resources with her other descendants. 'So if you think about the relatedness asymmetry in the human case,' Croft said, 'then evolution will favour females that compete more when young and help more when old.'

There is mixed evidence for this familial conflict in humans. A robust study using two hundred years of data from pre-industrial Finns offered support, whereas another smaller study on Norwegian women did not. 'It's hard to test in humans because we can't go back in evolutionary time. Which is why the killer whale system is so exciting as a way of testing these hypotheses,' Croft said. 'Who would have thought that we would find out so much about our own evolution by looking in the ocean at these toothed whales?'

Studying menopause in a six-tonne swimming torpedo with teeth is not without its challenges. One of the basic hurdles being, how on earth do you monitor the sex hormones that track an orca's reproductive senescence? Taking blood samples would be treacherous (for the scientist) and invasive (for the orca). A less intrusive, if smellier, alternative is to collect faecal samples. Which is how I came to be cruising the Salish Sea in search of orca poo on a sunny September afternoon with Dr Deborah A. Giles, director of research and science at Wild Orca and the southern residents' official scientific pooper-scooper. Giles has been studying the southern residents for the last ten years, knows them as individuals, and was my best ticket to an intimate encounter with an orca matriarch.

Giles plies her trade from San Juan Island, one of the hundreds of rugged landmasses left behind by ice age glaciers in the flooded fjord land that straddles Canada and the Pacific Northwest of America. Just a short hop on a seaplane from Seattle, the spectacular sixty-minute journey gave me a bird's eye view of the crazy cold-water currents and kelp forests that make the Salish Sea such a boon for marine life. The island's main town, Friday Harbor, was awash with orca imagery: wooden killer whales dived from streetlights, breached on walls as murals and waved to me as oven gloves from souvenir shop windows.

It took just twenty minutes to traverse the tiny island by car and meet Giles at Snug Harbor, where her modest speedboat was docked. She informed me that a shortage of hands meant I was being immediately promoted to spotting and scooping duty. A thousand questions sped through my mind. As a dog owner, I'm no stranger to picking up turds, but it was hard to imagine a bag big enough for this particular job. The ivy-green water looked deep and chilly – not to mention swimming with apex predators. Was diving involved? Giles handed me a large net and told me not to worry; whales do floaters, big ones.

'Poop is an absolute gold mine,' she told me. Scat samples enable her team to monitor not just the orcas' oestrogen levels, but also their stress and pregnancy hormones. They can tell what the whales are eating and check for the presence of parasites, bacteria, fungus and microplastics. Faecal samples provide a health check of not just the cetaceans of the Salish Sea, but their whole ecosystem.

First Giles has to find the faeces. Given the vastness of the ocean, even excrement from an animal as large as a whale is difficult to spot. Fortunately, Giles has help in the shape of Eba: a former street dog from Sacramento who's been rescued, rehomed and trained to sniff out whale scat.

Eba's twitching snout contains three hundred million olfactory receptors, which, compared to my paltry six million, means her scat-sniffing skills are around forty times better. She can smell whale scat from a nautical mile away and made the perfect poop-hunting partner. The little white rescue mutt was a ball of energy who clearly relished her renewed purpose as a conservation canine. 'Having a dog as your co-worker makes coming to work a hell of a lot easier. Plus, look at my office,' Giles said, gesturing to our silvery blue surroundings, sparkling in the low autumnal sun.

The first whales we encountered were a pair of humpbacks coasting the San Juan Channel – the rhythmic emergence of their four-metre-wide flukes gracefully hinting at their monumental thirty-tonne mass. These Brobdingnagian beasts are something of a conservation success story. Decimated by commercial whaling in the first half of the twentieth century, they've made an impressive comeback since the hunting of humpbacks was banned in 1966. Last year the local humpback ID catalogue had photos of one hundred individual tail flukes (the humpback equivalent of a fingerprint). This year it's four hundred.

Giles set course about fifty metres behind their disappearing tails in order to execute a 'distant poop follow'. Humpbacks feed on bait fish, as do Chinook salmon, so their excrement would tell a story relevant to the health of the southern residents. Our close proximity to the whales taught me something new too: whale breath smells really, really bad. When we found ourselves engulfed in a lingering cloud of stench, not unlike lifting the lid off a rubbish bin in summer, I assumed it meant our faecal quarry was close. But Giles soon set me straight: 'That's just breath. You think that's bad, you should smell minke – it makes you want to throw up.'

I counted my blessings and joined Eba in 'her office' at the front of the boat. Giles told me to scan the water for anything gelatinous. A fried-egg jellyfish and some blobs of decomposing eel grass raised the first few false alarms. Then I spotted a dinner-plate-sized dollop of something gooey and brown bobbing along the surface. We doubled back to give Giles the chance of closer inspection. 'I'm sorry to say that's boat bilge,' she said. It was excrement, yes, but human not whale.

Trawling the ocean for the right kind of turd may not be everyone's ideal job, but Giles wouldn't be doing anything else. When she was six years old she had a vivid dream about saving the southern residents. Back then, in the 1970s, it wasn't starvation and pollution that threatened these orcas, but kidnap. It was this population of killer whales that were brutally plundered by marine parks; nearly 40 per cent of the southern residents were snatched from the Salish Sea and imprisoned in aquariums for human entertainment.

'We've pretty much done everything we possibly can to decimate this population of animals, which makes me simultaneously incredibly sad and incredibly angry,' Giles told me, with evident emotion.

I was all too aware that I was joining Giles at a very dark time. In the eighteen months since Granny had died they'd lost a further seven southern residents, including two more post-menopausal matriarchs. Some of the whales were obviously emaciated – their bodies transformed from a big fleshy bullet to the deflated 'peanut head' associated with late-stage starvation. The population had hit a thirty-year low of seventy-three individuals, with the dead not being replaced fast enough to keep numbers up. Giles' faecal hormone study had shown that 70 per cent of pregnancies are failing thanks to nutritional stress, 23 per cent in late term.

The most heartbreaking loss was a newborn calf that had made global news after its mother, Tahlequah, carried her dead body around for seventeen days. The world's media speculated whether or not this young mother could be mourning; to Giles it was obvious. 'I find it insulting that we would assume she wasn't grieving,' she told me. 'These orcas are very much like us, but, quite honestly, I think they are better than us. They've got parts of their brains that we don't even have.'

Orca brains are a magnet for superlatives: vast and curiously complex, and as such not that easy for a human to wrap their comparatively limited grey matter around. Orca brains are the heaviest on the planet – around 7 kilos – and huge. You could comfortably fit five human brains inside an orca's, which is 2.6 times the volume you'd expect from a mammal their size – greater than the great apes'. The size of brain relative to body – the snappily titled encephalization quotient or EQ – is considered a rough proxy for intelligence. Humans have an EQ of around 7.4–7.8 and chimpanzees of around 2.2–2.5. Female orcas have an EQ of around 2.7, which is higher than a chimpanzee's, and also outdoes the males of their own species. Male orcas, with their bigger bodies, have an EQ of just 2.3. This disparity between the sexes flies in the face of Darwin's proclamation of male intellectual superiority, and is thought to be linked with the female orca's increased social and leadership skills, which require more cognitive power than the male needs.*

Size isn't everything, of course, but orcas have evolved proportionally more of the thinking part of the brain than humans have. Their cerebrum makes up 81.5 per cent of brain volume compared to our 72.6 per cent. Computational capacity is measured in size and surface area of the neocortex (the seat of complex thought) and the orca's is the most convoluted on the planet. If that wasn't enough to make you feel mentally inadequate, orcas also possess an enigmatic extra lobe of tissue that's snuggled in between their massively elaborate neocortex and the limbic system (where emotions are processed).

To understand what all these mind-boggling statistics actually

---

* Sperm whales show an even greater discrepancy of EQ between the sexes. The EQ of female sperm whales is more than double that of males (1.28 for females versus just 0.56 for males). This extraordinary degree of dimorphism is unique amongst mammals. And, as with the orcas, it is thought to be connected to the female's increased need for social intelligence. Male sperm whales are solitary creatures, whereas females live in large families in which social interactions and inter-individual communication are essential. When it is time for courtship, one can only imagine how limiting a female sperm whale must find conversation with the male.

mean I talked to Dr Lori Marino, who has devoted thirty years to the study of cetacean neuroanatomy and performed MRI scans on the brains of stranded orcas. She told me this so-called paralimbic lobe is only found in dolphins and whales. It provides dense connections between the two neighbouring areas of the brain and suggests that orcas could be processing emotions in a way we cannot comprehend.

'I think orcas experience a range of emotions from the joy that you saw [in Seattle] to despair,' she told me. 'I think it's likely there are dimensions to that emotional rainbow that we don't have and are hard for us to grasp.'

According to Marino, orcas have other parts of their brain involved in social awareness and communication that are also unusually complex. 'So what's interesting is that there are so many parts of their cerebrum that are more elaborate than primate brains, and these are the parts that do very interesting things – social cognition, awareness, problem solving – so the question then becomes, what is their psychology like?'

Marino thinks that orcas are emotionally sophisticated, lightning-quick thinkers with 'many more dimensions to their communication' than we have. They are one of the handful of animals that have passed the famous mirror test, paying attention to their own reflection in a way that suggests they have a sense of self. They are, however, far from selfish. Marino has speculated whether these 'socially complex brainiacs' might even have a distributed sense of self that's bound to the group, as well as the individual. This could explain the extraordinary levels of social cohesion that occur, even to their detriment. The reason the southern residents were so massively plundered by marine parks is that when one animal was caught, its family would remain by its side and could be added to the catch with tragic ease. 'They're perfectly capable of scooting it out of there,' Marino told me, 'but leaving the group is unthinkable for them.'

Aerial drone work by the Center for Whale Research is providing a fresh perspective on the intimacy of these social bonds. For the last forty years all orca researchers have been able to study is a parade of fins and the odd flash of body, now they're able to see what's going on beneath the surface. 'It's like lifting the lid on a fish tank and peering in for the first time,' Darren Croft told me. 'They've got a whole ocean to swim in and yet they're swimming not just next to each other, but touching each other.'

If you live in a vast, featureless three-dimensional space like the ocean, daily travelling great depths and distances, there is no such thing as a home to where you can retreat every day, to connect with loved ones and feel secure. Your family group *are* your home, your safe space and the key to your survival. So it pays for orcas to stay close and connected, in ways that we perhaps cannot comprehend.

They certainly display extraordinary levels of social support, including babysitting for each other's calves and caring for disabled individuals. Giles told me of a male orca from a transient mammal-eating pod, outside of the southern residents, that has scoliosis, yet is a thriving member of his family. 'They bring him food,' she told me. 'It's hard for him to keep up with them, but they loop back and bring him chunks of seal or whatever it is they've just killed. In a lot of human cultures they would leave that individual behind.'

I can't help but think about how many of our human leaders would benefit from a paralimbic lobe transplant – to become more like these wise and compassionate matriarchs, with their unfathomable emotions and supportive, socially inclusive society.

Not all orca females make great leaders, though. Giles told me that, just like us, orcas have different personalities, with some being more headstrong than others. 'We have "matriarchs of family groups" but they don't make good leaders. They go wherever the rest of their group goes. Then you have these others that clearly have a mind of their own and are directing the rest of their pod.'

The relationship between personality and leadership has been more conclusively probed in the Amboseli elephants, where behaviours are somewhat easier to observe than in a fast-moving submarine mammal. Dr Vicki Fishlock, the resident scientist at Amboseli, told me that personality differences play a major role with elephant matriarchs, although it's hard to quantify as families tend to perpetuate certain traits down their ancestral line, be they confident and curious or nervous and neophobic. A recent study by Fishlock's superiors, Cynthia Moss, who founded the Amboseli Project, and Phyllis Lee, professor of psychology at the University of Stirling, Scotland, analysed the matriarchs and found that leading the clan was less about being dominant and exerting power – less than for, say, an alpha male chimpanzee – and more about elevated levels of influence, knowledge and perception that earned the respect of the other elephants and gave them the confidence to follow.

Elephants, like orcas (and great apes, like us), have what's called a fission–fusion society, meaning their social lives are fluid. Group size isn't fixed, it is dynamic and can change by the hour as members split off and then rejoin one another. 'They're not told where to go, but this leadership is an attractive social centre,' Fishlock told me. 'The matriarch is the glue that holds everyone together.'

Losing this wise old female focal point shatters their social world, as Fishlock and her team discovered after Amboseli was hit by a horrific drought in 2009. It was the worst for several decades: rivers evaporated into the sky and grasslands shrivelled to dust. The Amboseli Project lost 20 per cent of their elephants. Old elephants are particularly sensitive to drought as their teeth are wearing out and they can't cope with eating the tough vegetation that survives without water. As a result the 2009 drought claimed 80 per cent of the Amboseli matriarchs over fifty. Amongst those lost was Echo, a legendary sixty-four-year-old female who had led her clan for almost four decades: a demise that has parallels with the southern residents losing Granny.

'Losing a matriarch affects everybody,' Fishlock told me. Most obviously, the clan are robbed of their library of ecological and social knowledge – the very thing they need to help them out of tough times. They no longer know who to look to for quick confident decisions, and that causes general confusion amongst the group. Equally damaging, however, is the social and emotional impact of bereavement.

'I see it as a trickle-down effect,' Fishlock told me. 'Grieving animals are not as responsive to others. So that has a knock-on effect on how closely they're bonded; they're a bit depressed, they don't spend as much time feeding and they don't tend to group needs.' A study on the effects of poaching on elephants in Mizumi, Tanzania, found that stress hormones were highest in groups that lost an old matriarch.

Fishlock thinks the impact of grief on such gregarious creatures is what leads to a higher proportion of family fission immediately after matriarch loss, which the Amboseli team observed after Echo's death.

Echo's sister Ella, aged forty-four, was the next oldest and could have taken charge but 'she couldn't be bothered with everyone else and just hived off with her family', Fishlock said. This left two females as potential matriarchs: Eudora, thirty-seven, and Enid, twenty-seven. Age, and the sagacity it conjures, generally trumps all in the elephant-matriarch stakes. But Eudora's temperament – 'she's a bit of a flake and personality-wise all over the place' – wasn't conducive to the top job. So, Echo's eldest daughter Enid became the matriarch even though she was ten years younger than Eudora: 'But that's pretty rare.'

It took the Amboseli elephants two years to figure out their new social structure and recover from the drought, but the group are now thriving. 'What's really cool is they've bounced back,' Fishlock told me. 'It looks like an intact age structure [one that includes both young and old members] is one of the key things that allows that to happen as it allows a new generation of leaders to emerge very naturally.'

The southern residents are still very much in the throes of their crisis. 'Time will tell what it really means to be losing all these older females. It's a scary thing to contemplate,' Giles told me.

They are certainly becoming increasingly fragmented. When Giles and I finally caught up with the southern residents, there were just two of twenty-three members foraging off the west coast of San Juan Island in a traditional salmon buffet spot that would once have drawn the whole pod. First I spotted the lofty onyx fin of Lobo, a nineteen-year-old male, surfacing so close to the boat my heart skipped a beat. Then I clocked his mother, Lea, swimming close by. At forty-two, she was probably on the cusp of her fertility and could, like me, be surfing a heated ride of middle-aged hormonal changes. I couldn't resist asking Giles if she thought that menopausal killer whales might also have killer moods and hot flashes during their change.

Orcas are basically 'an insulated sausage', she told me, making it hard to monitor their temperature. She could think of three incidents over the years when she'd noticed older females break away from their family and head off by themselves. 'That might be because they're cranky or they simply need some alone time, who knows,' she said. Comparing hormone levels would provide further clues. Hot sweats and raging moods are linked to a fall in oestrogen, which affects serotonin levels – a neurotransmitter associated with happiness. But these tests have yet to be done on the orca faecal samples.

Nevertheless, I felt an inspiring sense of communion with this killer whale. Lea (like me) was a social creature embarking on the next phase of her life. For her, the death of her ovaries heralded a rebirth of agency. She wasn't going to fade from society, but take centre stage – her aged insight commanding her clan's respect and propelling it forward. As one of the eight remaining post-menopausal females in the southern residents, might she even become the new Granny?

'Everyone is asking who is going to take over, but I don't think they even have that as a luxury of an issue,' Giles told me. She believes there simply isn't enough Chinook salmon to keep them together as a group and that's impacting the way they've traditionally lived their lives. 'It feels as if the entire fabric of their culture is unravelling.'

That culture is part of the problem. There is plenty of potential killer whale prey in the Salish Sea, but the southern residents have specialized in hunting salmon and won't chase anything else, even when they're starving to death. They share the same waters with a different eco-species of orca that catches marine mammals, and their population is booming. Occasionally one of the southern residents is seen toying with a seal pup, but much to their conservationist's frustration, they have yet to see them as food.

Most of the time we think of culture as conferring huge advantages, but the southern residents are teaching us the risk of cultural conservatism. In this rapidly changing world it pays to be opportunistic or you can end up in a dead end. What's urgently needed is for one of the emerging matriarchs to be an innovator, to stop playing with pinnipeds and instead take a big bite. That's how their culture will have started, many thousands of years ago, but who knows if such behavioural plasticity will happen in time to save these orcas.

Which means it falls to us humans, as the architects of all this change, to summon enough compassion in our paralimbic-lobeless brains to change *our* habits, before it's too late.

'If we humans don't make wholesale changes to the way we are doing business – from shipping to fisheries to toxicants in the water – then we are dooming this population. It makes us morally corrupt – we have done this to them,' Giles said, her voice shaking and barely holding back the tears. 'These whales are the canary in the coal mine for this region. They're telling us that things are wrong in the environment. If we save these whales we will have done a tremendous thing.'

CHAPTER TEN

Sisters are doing it for themselves: life without males

The Laysan albatross looks like a seagull with a serious steroid habit. They may be the most diminutive of the twenty-two albatross species, but their wingspan would nevertheless make the basketball giant LeBron James appear positively petite. The seabird's exceptional physique is adapted for dynamic soaring, a specialized form of flight that enables the albatross to cruise the blue planet with minimal effort, kept aloft by oceanic updraughts for thousands of kilometres whilst barely twitching a wing. Albatrosses can spend literally years at sea, their webbed feet never touching the ground, making these marathon mariners sacred to sailors, poets and myth-makers alike.

The need to reproduce brings a rude end to their lyrical nomadic existence, grounding them on remote rocky outcrops in raucous, overcrowded colonies. Laysan albatrosses favour the Pacific islands of Hawaii, where, after six months of solitude, they gather every November to mate and raise a single baby bird. This job cannot be done alone. Albatross chicks are uniquely slow maturing and take almost six months to leave the nest and become airborne themselves. During that time their parents must tag-team foraging missions as far north as Alaska, with one spending weeks at sea in search of squid to feed themselves and the growing demands of their chick, while the other sits tight at the nest, protecting their noisy but needy investment.

Such teamwork requires Herculean levels of trust, understanding and dedication, which is why the albatross has become a symbol of another extraordinary feat of long-distance endurance: monogamy. Albatrosses live for sixty to seventy years and typically mate with the same bird every year, for life. Their 'divorce rate', as biologists call it, is among the lowest of any bird. Praise for their devotion to long-term loyalty and nuclear family life has won support from far corners. When Laura Bush visited Hawaii in 2006, the Republican First Lady commended the albatross couples for making lifelong commitments to one another. What no one, least of all Bush, knew at the time was over a third of those committed couples were, to put it anthropomorphically, lesbians.

'You cannot assume that a couple are female and male,' Lindsay Young told me as we picked our way through a nesting Laysan colony in Hawaii.

Dr Young is the executive director of Pacific Rim Conservation and has been studying the Laysan albatross since 2003. Few people know these birds better, and I was lucky to accompany the Canadian biologist on her weekly census of the colony at Kaena Point, the most westerly tip of Oahu. This stunning finger of gently undulating sand bound together by a leafy tangle of native creepers emerges from the shadow of a jagged volcanic ridge masquerading as a dragon's back. It is wild and windswept and a world away from the Tiki bars and high-rise condos of Honolulu. As such, Kaena Point is a valuable preserve for a host of endangered seabirds – albatrosses, boobies and shearwaters. For the last twenty years these naive ground-nesting species have been protected from the feral cats, invasive rats and marauding off-roaders by the kind of massive metal fence that Donald Trump dreamed of. Safe inside, the Laysan colony has become a haven for seabirds and blossomed, unexpectedly, into what Young has described as 'the largest proportion of "homosexual animals" in the world'.

The albatrosses of Hawaii have been documented by biologists for over a century, but their unconventional coupling went undetected until 2008. It is easy to see why. As I stumbled after Young – tripped up by the countless concealed shearwater burrows that littered the colony like booby traps – I was eyeballed by an army of identical goose-sized birds. All were wearing the same fixed expression of bemused indignation courtesy of their facial feathers. Neither their bodies nor their behaviour betrayed their sex in any way.

I watched as birds enthusiastically courted one another with their ritualized dance in what Young's colleague dubbed 'the singletons' bar'. A handful of albatrosses would gather, bobbing their heads like solemn groovers at a silent disco. After a while a pair would sync up with one another and then up the ante to include ceremoniously sniffing their wing-pits, clacking bills, throwing their beaks in the air and plaintively 'mooing'.

Single birds will spend several years attending these albatross discos, assessing the talents of various dance partners before finally choosing their first mate, which, given the length and intensity of the resulting commitment, is wise. An albatross needs to find a partner with whom they can communicate and coordinate, and all this synchronized dancing seems to be part of the test. It was quite a performance – as passionate as any Argentinian tango. It seemed obvious to me which pairs had a good thing going on – you could almost feel the energy as the more synced duos vibed off one another. But there was no hint as to whether the couple in question were hetero- or homosexual. Not even to Young, who simply shrugged her shoulders when I asked.

There was, however, one big oval clue that made Young and fellow conservation biologist Brenda Zaun suspect there must be something special about these birds. A Laysan albatross is physically incapable of laying more than one egg per breeding season – it is just too much of an energetic investment for her to create any more. Yet numerous nests at Kaena Point contained a second egg, or had one lying just outside, as if it had rolled out.

Scientists call this phenomenon a 'supernormal clutch' and they had been periodically documented in Hawaii since 1919. Ornithologists came up with any number of unlikely excuses for these double eggs. One suggested that some females did after all have the oomph to squeeze out two, while another blamed other females for 'dumping' their eggs in these nests. This last theory was the idea of Harvey Fisher, a doyen of mid-century albatross science who also declared, with somewhat rash conservative zeal, that 'promiscuity, polygamy and polyandry are unknown in this species'.

It never occurred to such traditionalists to sex the super-clutch couples to check they were indeed heterosexual. But it did to Brenda Zaun, a biologist with the US Fish and Wildlife Service, who was studying a Laysan colony on the nearby island of Kauai. Zaun noticed that specific nests wound up with two eggs every year. Albatross tend to return to the same nests year on year, so the distribution of super-clutches couldn't be chalked down to random egg dumping, since it was happening repeatedly to the same couples. On a hunch, Zaun took feathers from a super-clutch pair and sent them to Lindsay Young, so she could isolate their DNA in her lab and genetically determine the birds' sexes.

When the results showed that both birds were female, Young assumed she must have made a mistake. 'My first reaction wasn't, "I've made this great discovery," it was, "I've really screwed up my lab work." '

So Young took samples from every super-clutch pair at Kaena Point. When they came back as all female, she thought she must have messed up her fieldwork instead. So, she went back to the colony and took more blood samples but got the same result: the couples were all female.

Her next reaction was, 'No one is going to believe me unless this is 100 per cent indisputable.' So, it took her about a year of testing lab results to make absolutely sure. In the end Young repeated her field sampling and lab work no less than four times before she was confident to go public. The final score was indeed astonishing: 39 of the 125 nests at Kaena Point since 2004 belonged to female–female pairs, including more than twenty nests in which she'd never noticed a super-clutch.

The data has been consistent from the start of their records in 2004 on Kaena Point – every year around a third of the couples are both female. Young figured that these females must somehow be finding opportunities to copulate with males but then choosing to shack up and incubate their eggs with another female.

But why?

It turns out these females are pioneers, in every sense of the word. The colony at Kaena Point is relatively new – Laysan albatross have only been nesting there since the area was protected from invasive predators and the ravages of off-roading vehicles in the late 1980s. It is being forged by offspring from large, congested colonies on other human-free Hawaiian Islands such as Laysan and Midway Atoll, which are home to over one million breeding adults.

It is the young female albatrosses that tend to be the adventurous ones, flying off from where they were born and striking out to pastures new. Young males are more likely to stay at home and begin breeding at their birth colony. Which leaves these fresh colonies like Kaena Point and the one on nearby Kauai with a shortage of males for the females to partner with. Since being a single parent is not an option for an albatross, these innovative females have adapted by soliciting a male from an existing couple as sperm donor, then partnering with another pioneering female for the hard work of raising the chick.

Although both females may lay a fertilized egg, only one egg will survive. Birds have a brood patch – a featherless area on their undercarriage which is used to regulate the temperature of the egg – but it's only big enough for one. So, one egg always ends up getting discarded. *Which* one seems to be totally random. It is highly unlikely the females are getting competitive and pushing each other's eggs out of the nest. Young is pretty certain the females have no idea which egg is theirs, they just seem imprinted to sit on something round. 'I've even seen one trying to incubate a volleyball,' she told me.

The reproductive success rate for a female partnering with another female is therefore, at best, half that of partnering with a male. But that's significantly better than not breeding at all. The main challenge these innovative all-female couples face is during the first few weeks after the eggs have been laid. In hetero couples the male takes the first sitting duty, so that the famished female can go off for three weeks and gorge herself on squid to replenish her energy stocks after giving birth. If both females lay an egg, one has to remain on the nest, literally starving. So, the rate of nest abandonment is higher amongst all-female pairs. But if they get over this hurdle, Young told me, the fledging rates for their chicks are just as high as for the hetero couples'.

'If they can hang on until the egg has hatched, they are just as good at raising the chicks,' she said.

Some of these females will partner with a female for one or two seasons and then switch to an available male. Given the preponderance of females, Lindsay thinks male Laysan albatrosses at Kaena Point get to be the choosy sex. Her data suggests they are attracted to females that are proven to be good mums, having already successfully raised a chick, regardless of partner. What's interesting is that this case of 'reversed' sexual selection doesn't result in increased female competition, as the theory would predict, but increased female cooperation, which would have left Darwin seriously scratching his beard.

Some female couples never switch to a male and remain in committed homosexual relationships for many years, if not for life. Young has no idea why, but it seems their union just works. Towards the end of our tour of the Kaena Point colony, Young introduced me to half of one such couple: bird number ninety-nine, or 'Gretzky'.

'It's actually a male name,' Young said, laughing and looking somewhat embarrassed. Naming animals is frowned upon by some scientists, and it's even worse if you get the sex wrong. Young met this bird and her partner before she knew about the same-sex couples, so she assumed she was male. The young biologist nicknamed the bird after her favourite Canadian ice hockey player, the square-jawed macho man Wayne Gretzky – aka 'the

greatest hockey player ever' – whose shirt number just happened
to be ninety-nine.

The sex may be amiss, but the champion's name seemed oddly
fitting to me. Gretzky and her female mate have been together since
Lindsay began monitoring the birds seventeen years ago, likely even
longer. In that time, they have successfully raised eight chicks
together and are grandparents to three more, placing them at the
top of the league of Kaena Point's most successful albatross couples, be they straight or same-sex.

So what was the secret to their long-term reproductive success?

For a start, I thought they'd clearly chosen some prime real estate
for their nest. Gretzky was sitting proud at the top of the colony
with a great sunset view, looking out over the rest of the settlement
and towards the thundering Pacific waves. More importantly, their
nest wasn't some ill-advised exposed dip in the sand like some, but
an impressive doughnut of mud and vegetation snuggled under the
leaves of a native Hawaiian naio bush. This dark-green dwarf shrub
would provide their chick with much-needed cover. Being coated in
downy grey feathers and unable to sweat makes cooking under the
hot Hawaiian sun a very real issue for newborn chicks. Their only
defence is to lie back in the nest and shove their big cumbersome
feet in the shade to try to cool down, a system which, as I sweated
like a pig in the eighty-degree afternoon heat, struck me as one that
evolution could do well to improve.

Chick number nine was due to hatch any day. As Young gently
encouraged Gretzky to stand up and expose her egg for the scientist to examine, the bird clacked her bill defensively at us. Young
was surprised to note there was still no sign of a crack in the egg.
Normally the more experienced parents hatch first and after a precise sixty-five days. This chick was late, as were all the others on the
colony this year, something Young presumed to be connected to
recent climate change, and a cause for concern.

As Young stepped back to make her notes, Gretzky bent down
over her egg and made a high-pitched barking sound. 'She's speaking to her chick,' Young told me. I imagined Gretzky encouraging

her chick to hurry up and hatch. And she may well have been doing so. Albatrosses have an impressive array of sounds and body movements to facilitate their communication. The chat starts young – even at this unborn stage, chick number nine could answer back to its mum.

Being a successful albatross parent is all about communication, coordination and cooperation. Gretzky and her mate have clearly got it nailed. When Gretzky's partner finally returns from her Alaskan fishing trip to relieve the hockey champ and let her feed, Young told me, the two will indulge in all the same lovey-dovey dance moves, nuzzling and preening that heterosexual albatross couples engage in.

'They do everything identically to the male–female pairs. There's no difference in their behaviour that we've been able to tell,' she told me.

These intimate physical gestures strengthen the birds' pair bond by stimulating the release of the 'cuddle hormone' oxytocin and various feel-good endorphins, in just the same way that kissing, caressing and sex does in humans. As well as promoting feelings of love, oxytocin lowers stress levels and encourages social recognition, which is all useful for finding your mate after weeks or months at sea and soothing the trauma of extreme hunger. Not to mention avoiding conflict with your neighbours in a crowded, competitive nesting environment.

Among birds, high levels of what scientists refer to rather clinically as 'allopreening' are associated with cooperative breeding and low divorce rates. Perhaps the secret ingredient to Gretzky and her partner's success is the strength of their physical intimacy, which has kept them loved-up all these years.

'Albatross are just like people,' Young admitted, in a rare unguarded moment of anthropomorphism. 'A good portion are monogamous and stay with the same mate for long periods, some of those socially monogamous couples cheat, some divorce . . . it's the whole spectrum.'

That spectrum now includes committed long-term same-sex

relationships that produce the next generation by acquiring sperm donations from otherwise betrothed males. As such, Young's research may have chipped away at the albatross' status as an icon of heterosexual monogamy. But that image was an unrealistic projection in the first place, imposed on these birds by the unnatural moral aspirations of Western religion.

What it suggests instead is something much more inspiring: the innate flexibility of sex roles in nature and the ability of animals to innovate their behaviour to succeed in challenging new social and ecological environments. This is going to be increasingly important as we head deeper into the eco-apocalypse. More than 65 per cent of Hawaiian seabird colonies are threatened by rising sea levels, as they nest on low-lying coral islands. Midway Atoll and Laysan Island, where 95 per cent of the Laysan albatross nest, along with other endangered boobies and shearwaters, are projected to have all but disappeared by the middle of this century.

So, these pioneering lesbians, forging new colonies on fresh higher ground, are literally preserving their species.

If science can remove its hetero-biased goggles for just a moment and take a fresh look at other species in which the sexes are identical and there is, perhaps, a shortage of males, we may find other examples of pioneering female cooperative breeders. An obscure and rarely cited academic paper from 1977 documenting 'supernormal clutches' and female–female 'homosexual pairs' of Western gulls, *Larus occidentalis*, nesting on Santa Barbara Island in California, indicates there are certainly more Sapphic seabird parents out there.

The authors, George L. Hunt and Molly Warner Hunt from the University of California, believed their study to be the first report of homosexual pairing in birds, although they were aware of reports of supernormal clutches in ring-billed gulls dating back to 1942. The Western gull female couples in the paper represented 14 per cent of the breeding pairs and were also thought to be an adaptation to a shortage of males on the island. There have subsequently been similar reports amongst roseate tern, *Sterna dougallii*, at Bird

Island, Massachusetts, and Dr Young is certain there will be many other examples out there, just waiting to be discovered.

The Laysan albatross isn't the only lesbian pioneer that lives in Hawaii. On my final day on Oahu, I raced across the island on a mad quest to meet a creature that's taken things one step further and dispensed with males altogether. Populations of the mourning gecko, *Lepidodactylus lugubris*, are all-female – there are no males in existence. They're a super-successful species that have colonized these islands without any male input at all, by cloning themselves.

A female-only race of self-replicating lizards sounds straight out of science fiction, and therefore impossible for me to resist. As soon as I heard of their existence on Oahu I fired off a bunch of urgent emails to various herpetologists. Several connections later, I ended up with a date on the other side of the island, with a woman called Amber who could apparently introduce me.

Amber turned out to be Dr Wright, assistant professor of ecology at the University of Hawaii. I found her, as promised in her email, mowing the long grass in a far corner of the university's agricultural research station. After unintentionally almost scaring her to death – she couldn't hear me approaching from behind thanks to the deafening growl of her mower – Dr Wright graciously offered to take me to my quarry.

The university research station is a lush slice of Hawaiian farming life – neat rows of local crops like taro, sugar cane and coffee arranged in small plots, overlooked by yet another verdant volcanic dragon-back ridge. My wild gecko chase concluded in the least salubrious corner of this agricultural idyll – the rubbish tip.

Geckos are nocturnal creatures that need to hide themselves away from potential predators during the day. They also have no eyelids, which I would imagine makes sleeping in the dazzling Hawaiian sun somewhat challenging. Wright lifted up a large discarded plastic fence panel and there they finally were, about half a dozen mourning geckos, rudely awakened and scurrying for cover.

The ecologist, a professional lizard wrangler, bent down and, without hesitation, gently grabbed one for closer inspection.

For such an extraordinary creature, this sci-fi species masquerades as exceptionally ordinary – an otherwise unremarkable little beige animal about four centimetres in length, not including her long fat tail. Like all geckos this extremity could be discarded and regrown if caught by a predator – a second sci-fi string to her bow. Her third superpower is the ability to scale any surface, however smooth, and casually stroll upside down across ceilings thanks to the adhesive pads on her feet. Wright encouraged me to feel the underside of her tiny little outstretched hand. It was indeed sticky. Not because of any glue-like substance, but nano-scale fibrils that exchange electrical charges with whatever they touch, causing them to stick, a system so effective that NASA are co-opting it to develop 'astronaut anchors' that allow robots to adhere to the exterior of space stations when conducting repairs.

Wright turned the lizard over, and through the semi-translucent skin on her tummy I could clearly see two white blobs – about the size of Tic Tacs – nestled in her groin. 'Those are her eggs,' Wright told me. Unlike most other animals, this little lady lizard's eggs would not need sperm to complete them. All she had to do was find a safe spot to lay her clutch, and they would hatch into perfect little replicas of their mother, without any male assistance.

The gecko looked at me and licked her eyeball, the only way for her to moisten her golden eyes, without the luxury of eyelids to lubricate them by blinking. She wore an enigmatic smile. With zero facial muscles it was a fixed grin that betrayed no real emotion, but seemed fitting, nonetheless. I smiled back.

This tiny lizard has evolved a cunning approach to the stressful business of reproduction. Not for her the trials of hunting down a suitable member of the opposite sex and making herself attractive to him. She's swerved that energy-consuming ordeal. She need not summon the additional energy required to copulate successfully either. Or run the risk of being eaten in flagrante – sex can be an understandably distracting affair for the participants involved and their rapturous movements make them vulnerable to the eyes of passing predators.

No, this little mourning gecko was a one-lady lizard-making machine, capable of creating up to three hundred clones of herself during her short five-year life.

This shouldn't, of course, be possible. Egg and sperm cells are created through a process called meiosis, whereby an embryonic reproductive cell's chromosomes are duplicated before the cell divides twice. This produces four daughter cells, each with half the DNA of the original. It means that egg cells only contain half the total number of chromosomes that most other cells in the body do. It's their union with sperm, which are also genetically half-cocked, that restores the full balance of chromosomes, ready for the next generation.

The mourning gecko has somehow evolved to cheat this fundamental process. She joins a very exclusive female-only club of around one hundred known vertebrates – an assortment of fish, lizards and amphibians – and a host of spineless wonders only really visible to microscope owners, but whose sex lives (or lack thereof) are exploding evolutionary paradigms left, right and centre.

Their remarkable asexual skill is known as parthenogenesis, from the Greek meaning 'virgin' and 'birth'. It has enabled the mourning gecko to colonize not just the Hawaiian Islands but Sri Lanka, India, Japan, Malaysia, Papua New Guinea, Fiji, Australia, Mexico, Brazil, Colombia, Chile and many, many more places. In fact, if you visit pretty much any warm coastal region in the Pacific or Indian Ocean, the mourning gecko will likely be there, hanging out by the outside light of an evening, kindly consuming the mosquitoes for you and 'chirping' cheerily.

It is truly mind-boggling to think, but the global population of *Lepidodactylus lugubris*, which must be many, many millions, are all clones of just a handful of original mothers. Their global reach and resilience as a species beg the question, why bother with sex at all?

This has been dubbed the 'queen of questions', and is, perhaps, the greatest riddle in evolutionary biology. Sex, you see, is expensive. As well as all the inconvenience of finding and attracting a mate, it cuts your reproductive potential in half, since only one sex can actually produce eggs.

All-female species can proliferate at twice the rate of a sexual species. This makes them excellent colonizers of new territory,* since they're specialized for quick growth and fast dispersal. The mourning gecko arrived on Hawaii with the Polynesians in AD 400 and was ferried about the Pacific on WWII ships. She's managed to take hold pretty much everywhere we've transported her. Like many other unisexual species, including the marbled crayfish and Antarctic midge (*Eretmoptera murphyi*), she's considered a 'weed species' or 'alien invader' – a pejorative and highly subjective label that's somewhat ironic in young volcanic islands like Hawaii where little life, least of all human, can claim to be truly native.

These unisexual 'weed' species are explosive opportunists whose aggressive proliferation can muscle out slower-reproducing sexual competition. Yet despite the economic advantages of parthenogenesis, all-female species are very much in the minority. Nature remains addicted to sex.

The problem with cloning is all offspring are genetically identical to their mother. Essentially, it's the ultimate form of inbreeding – there is no way to create genetic diversity, except for the occasional copying mistake during meiosis. So, animals that clone themselves leave their lineages vulnerable to parasites, disease and environmental fluctuations, which they lack the genetic variety to counter.

---

*   The potential for doubling production in an all-female species hasn't escaped the attention of agricultural scientists looking to maximize profit. Chickens and turkeys are just two farming staples that have been selectively bred to create all-female strains. Dolly the sheep is another example. Her revolutionary cloning from a mammary gland cell in Scotland in 1996 led to the ewe being named after the voluptuous Ms Parton and worldwide fame. Less well known was the venal reason behind her creation, which reflected a fifty-year aspiration to increase agricultural mammal production by taking the sex out of reproduction.

If environments were pestilence-free and fixed, males would be somewhat superfluous. But fortunately for them, the world is ever-changing, and we need sex to shuffle the genetic deck and maintain diversity. The benefit of sexual reproduction is survival over evolutionary time, which is why the most successful 'weed' species make use of both modes of reproduction.

Gardeners will be painfully familiar with one such self-cloning creature, for her ability to ravage their tomatoes, amongst other things. I'm talking, of course, about the aphid. The four thousand or so species of this small sap-drinking insect are despised for sucking the life out of crops and spreading disease. They are also, quite possibly, the virtuosos of the cloning world.

At the start of summer, a single female will give birth to 50–100 whole females, each already pregnant with embryos developing inside her. Like little plump green Russian nesting dolls this telescoping of generations shortens the nymphs' maturing time to just ten days and enables aphids to boom exponentially. Some species of cabbage aphid, for instance, like *Brevicoryne brassicae*, can produce up to forty-one generations in a single season. So, one female hatched at the start of summer could theoretically produce hundreds of billions of descendants, were it not for ladybirds gobbling them up.*

---

* The celebrated eighteenth-century French naturalist René Réaumur calculated that if all the descendants of a single aphid survived during the summer and were arranged into a French military formation, four abreast, their line would extend for 27,950 miles, which exceeds the circumference of the earth at the equator. It's fair to say Réaumur was pretty obsessed with aphids. The avid entomologist wrote with great passion in his bestselling *History of Insects* about how he could not, despite many, many hours searching, find a male aphid nor indeed a pair engaged in the act of 'coupling'. He even tried watching a virgin female for days to catch her in the act before giving birth but failed. Réaumur's abortive experiment was, however, picked up by a budding young scientist named Charles Bonnet who took on the aphid experiment with unflinching doggedness. He imprisoned a virgin aphid in a jar and watched her for her entire thirty-three-day life, around the clock from four in the morning to ten at night. He was certain

In the autumn, once the damage is done and their numbers are legion, the females reproduce sexually with male aphids, and lay down eggs with the necessary genetic variety to cope with whatever the following year throws at them. It's a bulletproof system that's turned them into the gardener's greatest nemesis. In the end, the aphid always wins.

So, females need males to maintain genetic diversity. The males also have one other essential job. Sex prevents the accumulation of harmful mutations that naturally occur during meiosis. In sexual species, these copying mistakes get switched out when sperm and egg combine by the healthy genes on the opposite sex cell. Asexual species don't have this luxury and are, instead, doomed to endless replication, gathering in number until they result in what's known in genetic circles as 'mutational meltdown'.

This is as bad as it sounds and the reason why exclusively asexual species have a reputation for being short-lived compared to those that at least dabble in sex, earning these profligate all-female species the negative branding of 'evolutionary dead ends' on the tree of life.

Your average sexual species is around one or two million years old. Asexual species, on the other hand, rarely get to celebrate beyond their one hundred thousandth birthday. Or, at least, that's the theory. The trouble is there are a bunch of all-female species that have flagrantly ignored their shelf-life predictions. These 'evolutionary scoundrels', as the legendary biologist John Maynard Smith has described them, have caused no end of scientific fluster, calling into question the whole question of sex.

You may feel that you've weathered some overly long sexual droughts, but they are nothing compared to the bdelloid rotifer, whose commitment to celibacy is unmatched in the animal kingdom. These microscopic relatives of the flatworm haven't had so

---

she never had sex yet she produced ninety-five young. His painstaking dedication to aphid observation paid off. In 1740 Bonnet became the first person to disprove the universality of sex in nature and declare the existence of 'virgin births'.

much as a sniff of sex for some eighty million years. All four hundred and fifty species of this class of rotifer are female. They make their home in brackish water, such as puddles and sewage treatment tanks, which wouldn't be enticing on any Tinder profile. But these self-replicating sisters don't care, because they have cracked the conundrum of how to survive natural selection and evolve without sex.

The reason for the evolutionary longevity of the bdelloid (pronounced with a silent b) is an area of continuing investigation and furious deliberation among scientists. However, one of the secrets to their success seems to be that they 'steal' genes from other lifeforms, possibly through the stuff that they eat.

The bdelloid diet isn't exactly enviable, which given their living conditions is hardly big news. They mostly survive on 'organic detritus' (don't ask), dead bacteria, algae and protozoans – pretty much anything they can fit in their mouth. Some scientists think that the bdelloids can extract DNA from their dinner and spruce up their own genome by a process called 'horizontal gene transfer'. Studies have shown that up to 10 per cent of the active genes in these rotifers could be pirated from other species. All told, the bdelloids appear to have adopted a Frankenstein collage of foreign DNA from more than five hundred different other species. Whether that's through ingestion or not is up for debate, but these pilfered genes provide the bdelloids with much-needed genetic variation in the absence of sex. They could also be the reason for another bdelloid superpower: their epic resistance to drought and radiation.

In nature's edition of the TV reality show *Survivor*, the bdelloid rotifers would probably be the last animals standing. They can survive several years of desiccation and high blasts of radiation. As far as we know, they are the most radiation resistant creature on the planet, beating even the famously indestructible tardigrade in the hard-nut stakes.* Bdelloids can withstand being blasted with one

---

* The tardigrades or 'water bears' are microscopic creatures that can enter a dormant, dried-out state where they can withstand extreme heat, temperatures

hundred times the radiation that would melt you or me into a hot puddle, and *still* turn around and give birth to a handful of healthy daughters.

The bdelloids are able to survive and produce viable replicas of themselves because their mosaic of stolen genes code for enzymes that give them the remarkable ability to repair their shredded DNA. The temporary water sources they call home routinely dry out, leaving the bdelloids essentially mummified for potentially years while they wait for their next rain. Such dehydration has the same damaging effect on their DNA as radiation (if less extreme); their stolen genetic repair kit helps them patch things back up. As an added bonus it's likely this is the bdelloids' secret to surviving millennia without males; the process of stealing genes and reconstructing their genome looks like it provides the bdelloids with the same evolutionary benefits as sex.

This is huge news. Until the bdelloid rotifer came along, sex had cornered the market on shuffling the genome. Now we've discovered there's at least one other way (males, take note).

It turns out many other all-female species have been hiding their age and are in better evolutionary health than would be expected, having evolved their own sneaky ways of mitigating the threats of mutational meltdown and genetic uniformity. In 2018, the unisexual *Ambystoma* salamander species was discovered to be five million years old. Her ticket to evolutionary old age is 'kleptogenesis'. She occasionally steals sperm from closely related species, which are used to stimulate but not fertilize her egg. Every once in a millennium, however, she incorporates bits of her stored 'stolen' sperm into her genome in order to maintain diversity.

These are exciting times. The discovery of such genetic jiggery-pokery is blurring the lines between what is a sexual and what is an asexual species. Even challenging the very fundamental question of

---

close to absolute zero, poisonous gases and extreme radiation. They've even survived the open vacuum of space. But even these hard cases have been sterilized by the 500–1,000 Gray doses of radiation tolerated by the bdelloids.

what a species is. These all-female pioneers are challenging hetero-normative assumptions about the ubiquity of sex and opening up new areas of enquiry for genetics and evolutionary biology. There are, for instance, surprising physical and behavioural differences amongst these clonal lineages, suggesting that epigenetics – differences in the way genes are expressed – might play a part and foster some variation, even in genetically identical lines.

As for the mourning gecko, the secret to her parthenogenesis turns out to be that she's a hybrid of two closely related species. Hybrids are generally considered sterile – they can't reproduce. But our biologically rebellious mourning gecko didn't read the memo and, instead, this hybridization sparked her switch to cloning.

The mourning gecko's hybrid parents were genetically distant enough that their chromosome sets are mismatched, but close enough to still undergo a form of meiosis. This hybrid sweet spot seems to be the trigger for parthenogenesis in many other known unisexual vertebrates. The resulting clonal daughters still lack the genetic mixing that gives sexual reproducers a steady flow of vari-ation to fight change. But there seems to be some mixing that occurs between these diverse hybrid chromosome sets during their idiosyncratic meiosis which helps limit the impact of inbreeding over time. Some clones even have an additional set of chromo-somes, implying the hybrids are also hybridizing. This may confer further genetic variation and longevity. It's all very new. Like the Nexus-6 replicants in the original *Blade Runner* movie, these clones have a shelf life that's greater than expected, but who knows how long.

The most curious thing about these little unisexual lizards is that even though they don't need to have sex to survive, they haven't quite been able to kick the habit. An obscure paper by an Israeli professor, Dr Yehudah L. Werner, published in Berlin forty years ago, describes the 'homosexual behaviour' of these geckos on Oahu. Females were observed wrestling and mounting one another, with one lizard playing the 'male' role and the other the 'female'.

This isn't the only report of lesbian lizards out there. All-female desert grassland whiptails also indulge in lengthy courtship and 'mating' behaviour before laying eggs, even adopting the so-called 'doughnut' sexual position favoured by their closely related sexual cousins. Initially these 'pseudo-copulations' were assumed to be simply vestigial – a hangover from their sexual past. But why the gendered sexual role play?

In many sexual lizard species courtship and copulation are necessary to stimulate egg production. With no males available for ovary-arousal duty, those wily whiptail females have become gender fluid.

When two female whiptail lizards are housed together in a terrarium they take turns switching sex roles each monthly egg cycle. One month one lizard gets to be the mounting 'male', the next month she plays the mounted 'female'. This curious case of role play etiquette turns out to be hormonally driven. Once placed in the same enclosure, two females' cycles sync up as the opposite of one another – so roughly every two weeks one lizard is ovulating. High progesterone levels post-ovulation trigger the females to act like males and start mounting. This complex behaviour is controlled by neural circuitry that is typically switched on or off in males by testosterone. But in the unisexual whiptails it's also activated by progesterone, allowing these all-female lizards to switch sex roles and maximize their fertility. Studies have confirmed that whiptails that don't get the chance to fake mate don't lay as many eggs.

These unisexual whiptails are outliving their evolutionary shelf life and maximizing their reproductive fitness, all in the absence of males. But is their sci-fi all-female society a utopia or a dystopia?

A fascinating experiment in the 1980s by Dr Beth Leuck of the University of Oklahoma took several species of whiptail – both asexual and sexual – and housed them in an enclosure for close observation. The all-female species behaved very differently to their sexual cousins. Same-sex couples were much more likely to shack up and share burrows than their sexual relatives, who tended to sleep alone at night. The sexual societies also showed four times more aggression against one another than the asexual species. Sexual species got into more fights, did more chasing as they tried to steal one another's food and had a stronger dominance hierarchy.

The all-female clones share 100 per cent of their DNA with each other, making them significantly more closely related to one another than the sexual whiptails. This kinship could account for their increased cooperation. Leuck observed that males were the instigators of much of the aggression, suggesting that, in whiptails at least, an absence of males results in a more tolerant society. All of which kind of makes me want to be reincarnated as a unisexual whiptail lizard.

The whiptails are certainly flying the flag for a life without males. The success of these all-female species demonstrates a new dimension in the battle of the sexes, with males fighting not just to fertilize eggs, but to simply exist. All-female societies are twice as productive without the dead weight of males, whose offering of genetic diversity is now understood to be less crucial than previously assumed. Recent mathematical models have revealed that evolution need not favour sexual reproduction, even when it does increase variability that is beneficial, and so, the question of sex remains an enigma.

But what about more socially complex societies, where males are more than just sperm donors or egg-stimulators – surely they're safe? Well, a recent discovery in Japan has suggested perhaps not.

In 2018 Toshihisa Yashiro of Kyoto University reported finding the first all-female termite society. All-female lineages have previously been documented in a few ant and honey bee species, but their colonies are already dominated by queens and female workers. Yashiro's finding is major because termite colonies are created by both a queen and a king, which reproduce sexually to produce male and female offspring. In traditional mixed-sex mounds, both sexes were thought to play vital roles in maintaining their complex society. So, the loss of males would therefore be genetically and socially significant.

From a human perspective, termites are the unloved outcasts of the social insect world. Whereas bees are praised for their pollination skills and ants are lauded for their industry, termites are an affront to human civilization, chomping their way through everything we hold dear: our libraries, our homes, even our cash – in 2011 an errant gang of termites burrowed into an Indian bank and ate $220,000 in bank notes.

Termites are the original 'anti-capitalist anarchists' and, frankly, deserve more of our respect. Termites have been doing astonishing things since the time of dinosaurs, maintaining complex societies with divisions of labour, farming fungus, converting cellulose to sugar and constructing vast skyscrapers complete with their own air-conditioning system. And now we can add overthrowing the patriarchy to that list.

Dr Yashiro's team collected seventy-four mature colonies of *Glyptotermes nakajimai* from fifteen sites in Japan. Thirty-seven of the colonies were exclusively female, while the rest were mixed-sex. The all-female colonies had an extra chromosome compared with the sexual ones, suggesting the two groups may be diverging into different species, having originally split some fourteen million years ago.

Their social make-up was different too. In the sexual colonies a king and a queen produce male and female workers that perform various roles from nursemaid to armed guard and, later in life, often go on to reproduce themselves. The unisex colonies are ruled by a large number of cooperating queens. They also have fewer soldiers, leading Dr Yashiro to presume that the all-female army might even be more efficient than the mixed-sex one.

In a previous study Dr Yashiro discovered female termites controlling the switch to asexual reproduction by literally 'closing the sperm gates' and sealing up the sperm storage organs (spermathecae) in their abdomen so that males could no longer physically inseminate them. The queens in question went on to produce genetically identical daughters by cloning instead.

It is possible that *Glyptotermes* also actively signed the death warrant of their males by shutting out their sperm from their spermathecae. This decidedly sci-fi scenario is unlikely to be under the queen's conscious control but, nevertheless, Yashiro concludes that his revolutionary all-female termite society is the first to provide evidence that 'males are dispensable in advanced animal societies in which they previously played an active social role'.

Sex certainly seems to be losing its evolutionary edge. This may not be great news for males, but it could spell salvation for a wide range of critically endangered species harbouring a latent skill for cloning.

In recent years, virgin births have been cropping up in all sorts of unexpected species and places – for instance, sharks in *Nebraska*. This land-locked Republican Midwestern state isn't exactly known for its ocean life, miracles or indeed female emancipation. But a species of hammerhead shark changed all that when she gave birth in a zoo aquarium in Omaha, surprising everyone. Her tank-mates were two other female hammerhead sharks and an assortment of rays, so who (or indeed where) was the daddy?

Female sharks have the ability to store sperm for months, if not years, so it was assumed this shark, let's call her Mary, must have mated prior to being taken captive. The aquarium soap saga continued when the pup was killed by a stingray after just a few days of life. The unfortunate tragedy did at least give researchers the opportunity to do genetic analysis, which proved there was no 'DNA of male origin' in the baby shark. Gifting local papers with a unique opportunity to compare a man-sized shark with the mother of Jesus.

Instead of being fertilized by a male's sperm, the female shark's own genetic material combined during meiosis. A cell called the secondary oocyte, which contains half the female chromosomes and normally becomes the egg, fused with another cell called the secondary polar body, which contains the identical genetic material. Together these haploid cells made the full diploid set of DNA necessary to create a new hammerhead shark. Nothing unusual had happened to the shark, so the trigger for this event was unknown at the time.

The shark discovery proved to be another paradigm shifter. Parthenogenesis was otherwise unknown amongst the cartilaginous fishes, the primitive group to which sharks belong. They now join the bony fish, amphibians, reptiles and birds on the asexual bandwagon. Such prevalence suggests the process has ancient roots in the vertebrate line.

In the last few years there have been a flurry of similar 'virgin births' in zoos around the world to an unholy cast of characters. A blacktip shark, Komodo dragon and a six-metre reticulated python called Thelma all recently cloned themselves in captivity by similar means to the hammerhead.

Parthenogenesis appears to be a last-resort tactic for these zoo animals that don't have the opportunity to reproduce sexually. Perhaps, hundreds of millions of years ago, cloning evolved as a handy alternative sexual strategy for ancient vertebrates in disparate populations struggling to hook up.

With environments becoming increasingly fractured and with so many species in catastrophic decline, finding a viable sexual partner is likely to become increasingly hard. Females that can fall back on the ancient art of cloning might just be what's needed to help species weather tough times.

The sawfish, a type of ray with what looks like a chainsaw attached to its face, has recently been documented doing just this. Native to the rivers of west Florida, it is one of the world's strangest, and most critically endangered, sharks – sawfish numbers have declined to 1–5 per cent of their original population size. In 2015 researchers at Stony Brook University analysed telltale markers called microsatellites in 190 sawfish that reveal how related their parents are. In seven fish, the markers suggested their parents were identical to them. Which can mean only one thing: female sawfish had started to clone themselves.

This is the first ever documentation of this type of parthenogenesis in a shark, or indeed any kind of vertebrate, in the wild. It signals something horrifying – a tragic tipping point for a species on the cusp of extinction. But the discovery of these pioneering parthenogenetic females also provides me with a chink of hope. As a short-term strategy, cloning could maintain lineages through periods of isolation, with the option of reverting to sexual reproduction once a suitable male becomes available.

If you trust the maths, this might not even harm the gene pool. Recent models have shown that good mutations can actually spread almost as fast in a population that is primarily parthenogenic. The benefits of sex seem to top out if it happens just once every ten or twenty generations; as one recent paper put it, sexually reproducing just 5–10 per cent of the time is enough to get the same genetic advantages as doing it every time.

So females with the ability to switch to cloning and boost population numbers above critical levels may be just what's needed to save a species from extinction. In the case of the sawfish, conservationists are seeing some recovery, which could be down to these sexually innovative parthenogenic females.

The one problem with this optimistic scenario is us. As I sit here writing this book, it feels like the end of life as we know it. We are in the grip of a global pandemic, fires are destroying the Amazon and Australia, and unprecedented storms are whipping America, Asia and Europe. Climate change is very real indeed and transforming the planet so swiftly in places that even healthy-sized sexual populations will struggle to adapt fast enough. Our species needs to make radical changes – both individually and on a grand scale – and fast if we are to halt our rampant destruction of this planet and allow ecosystems to recover.

It doesn't take fancy mathematical models to show that if there is nowhere for a species to live, they are indeed doomed, however they reproduce. Parthenogenesis is a safety net only available for certain species and it is only females that have the power to clone.*

---

* Arrhenotoky is a special kind of cloning that produces males, but those males cannot then clone themselves – only females retain this privilege. Most bees, ants

These special females are likely to become increasingly important. If we keep on with our path of war and destruction, the future will be most definitely female: only the bdelloid rotifer will be left standing.

The only major vertebrate group that's still unknown to clone naturally are the mammals. Parthenogenesis has been induced in the lab,* but there is no known example of a mammal having a virgin birth, either in captivity or the wild. Fundamental aspects of mammal biology make it unlikely we ever will find any mammal self-replicating in nature. So, it looks like men can sleep safe (for the time being) and humans are going to have to stick to the messy business of sex. Which, as we are the architects of all this destruction and the ultimate 'weed' species, is for the best. The idea of humans reproducing like aphids is a truly terrifying scenario – and most certainly not what this planet needs right now.

---

and wasps reproduce in this way – females are produced from fertilized eggs and males from unfertilized eggs. Komodo dragons are one of a few known vertebrates known to also produce males asexually due to their unique ZW sex-determination system – males are ZZ and females ZW. A virgin birth in a Komodo female produces only one viable sex – a male with a single Z. I met one such reptilian Jesus, named Ganus, at ZSL London Zoo. His captive mother Flora surprised zoo staff by not only reproducing asexually but by having a son as a result.

* In the 1930s Gregory Goodwin Pincus, co-inventor of the contraceptive pill, claimed to have created a 'fatherless' rabbit by treating an egg with saline, hormones and heat in his lab. The resulting bunny clone became a big cover star. But when others failed to induce what the inventor had branded 'pincogenesis' by using the same recipe, doubt was cast on the validity of his experiment. Many decades later, in 2004, a Japanese lab produced a female mouse by artificial parthenogenesis. The mouse lived and went on to produce offspring.

CHAPTER ELEVEN

Beyond the binary: evolution's rainbow

The universe is not only queerer than we suppose, it is queerer than we can suppose.

J. B. S. Haldane, 1928

I'd like to begin this final chapter with an examination of Darwin's barnacles, as they have much to divulge about the great man and his thoughts about sex. Darwin may be famous for his finches, whose subtle beak variations between neighbouring Galapagos islands helped inspire his theory of evolution by natural selection. But Darwin's fancy for finches was nothing compared to his fixation with barnacles. This humble group of gritty crustaceans, commonly found encrusting rocks at low tide, latched on to Darwin's heart as a young man and developed into a lifelong passion.

The obsession began with Darwin cataloguing the barnacles he collected on his five-year global odyssey on board the *Beagle* in the early 1830s. Word soon got out to fellow zoologists and before too long Darwin's home in Kent was inundated with *Cirripedia* specimens from all across the world. From 1846 to 1854 Darwin documented his salty gifts with a zeal few had shown for barnacles before, or indeed since. He was so involved in his work that when one of his sons visited the house of a gentleman neighbour he is reported to have asked, 'Where does he do his barnacles?', as if studying barnacles was every father's occupation.

This labour of love proved so all-consuming it delayed the publication of Darwin's book on the origin of species by many years while he painstakingly produced four exhaustive volumes on the world's barnacles, both extant and extinct. They may not sound it, but for the few that have ventured to seek out these obscure barnacle monographs, they make for a surprisingly illuminating read.

Darwin's diligence was rewarded with a number of key discoveries. Barnacles had previously been classified as molluscs for their resemblance to limpets. Darwin established that they are in fact part of the same group as crabs and lobsters, only they have sacrificed their mobility for increased home security. The free-swimming larva of a barnacle cements itself to a rock by its head and acquires calcified protective plates. It then lives out a secure existence, catching food and filtering oxygen by waving its feathery legs about through a hatch in its armour.

The barnacle's sedentary life is a boon for personal safety, but less so for sex: locating a mate isn't exactly easy when you're glued to a rock. Darwin discovered the barnacle's secret weapon is an extravagant penis, the longest relative to body size in the animal kingdom. Darwin's customary functional prose takes an almost giddy turn as he describes how the barnacle's 'prosciform penis' is 'wonderfully developed' and 'lies coiled up, like a great worm' but 'when fully extended, it must equal between eight and nine times the entire length of the animal!'.

Such frivolous length is purely functional. It enables the barnacle to cruise the neighbourhood for sex, whilst remaining cemented to the spot head first. Like a sort of X-rated Mr Tickle. It helps that most barnacles are simultaneous hermaphrodites. Every individual has both male and female reproductive organs so they can fertilize and be fertilized by all of their neighbours. And if there's no one else within reach, as a last resort the barnacle can recall their roving inseminator and fertilize themselves.

In 1848 Darwin stumbled upon a specimen with no penis at all. Not only that, it appeared to be infested with tiny parasites. Darwin had been picking them off and throwing them away when he realized his mistake. The barnacle in question was a female and the microscopic 'parasites' were the male of the species, albeit somewhat abridged: mouthless, stomachless and short-lived.

In a private letter to colleague J. S. Henslow in 1848, Darwin described the pitiful locked-in life of these so-called 'complemental males' with palpable empathy. By this stage of his life Darwin was plagued by chronic ill health. His globe-trotting days were but distant memories and he remained grounded in Kent, a semi-invalid. The limited existence of these withered barnacle males, as little more than incarcerated inseminators, perhaps reminded him a little too closely of his own as a housebound father of ten. These 'mere bags of spermatozoa', he opined, are 'half embedded in the flesh of their wives', leaving them doomed to 'pass their whole lives & . . . never leave'.

One month later Darwin made an even more curious find: a closely related species of barnacle that was hermaphrodite, but *in addition* it had these tiny 'complemental males' embedded in its body. Darwin speculated that these individuals represented a transition in the evolution from hermaphrodite to separate sexed barnacles – a sort of missing link for sexual differentiation.

In a letter to his friend and intellectual sounding board the respected botanist Joseph Hooker, Darwin ventures how 'an hermaphrodite species must pass into a bisexual [i.e. separate sexed] species by insensibly small stages; and here we have it, for the male organs in the hermaphrodite are beginning to fail, and independent males ready formed'.

Darwin considered this curious barnacle as further evidence for his big 'species theory' that he was busy developing (eventually known as his theory of evolution by natural selection). Darwin's idea that all life evolved from a common ancestor, as opposed to having been created by God, was heretical enough. The suggestion that sexes can transform over time would have been a truly outrageous statement, even amongst lowly crustacea. Darwin signed off to Hooker admitting as much, but unable to contain his excitement at his discovery nevertheless: 'I can hardly explain what I mean, and you will perhaps wish my barnacles and species theory al Diabolo [to the devil] together. But I don't care what you say, my species theory is all gospel.'

Despite the sacrilegious sexuality of these tiny crustaceans, Darwin was clearly filled with wonder. What's fascinating to me is how these private letters and little-known early monographs contrast with Darwin's later, high-profile published work edited by his prudish daughter. Barnacles and their prodigal penises are conspicuously absent from *The Descent of Man, and Selection in Relation to Sex*. But in these confidential missives Darwin was at liberty to marvel at the novel sexual set-up of his beloved *Cirripedia* without fear of public controversy, something he worked hard to avoid. There's no sign of the rigid binary Victorian stereotypes seared into his theory of sexual selection. Instead we simply see the genius of Darwin's curiosity exploring the spectrum of sexual expression without fear of the Church, the establishment or his daughter's red pen.

In another of his private barnacle letters (this time to Charles Lyell in 1849) Darwin signs off by eulogizing, 'Truly the schemes & wonders of Nature are illimitable.'

Truly they are. A century and a half later, state-of-the-art research using DNA markers has confirmed that Darwin was indeed right. Barnacles display a rich diversity of sexual systems – from hermaphrodite to separate sexes to a mixture of both – allowing scientists an exceptional opportunity to study evolution in motion.

Barnacles are masters of hedging their sexual bets. Their ability to adapt their sexual system according to the environment or social situation they land in produces a spectrum of procreative possibility in adulthood. Dwarf males, for example, may or may not develop ovaries depending on whether they land on or near a female. Classifying them as strictly male is therefore tricky. Instead many are considered an indistinct sex better described by modern science as 'potential hermaphrodite' that 'emphasizes male function'. In some cases the boundary between a hermaphrodite individual and a female, or dwarf male, is so blurred their sexual expression is considered more of a continuum than a distinct classification of sex.

The barnacle's rapid evolution from one reproductive system to another reveals the surprising flexibility of sex and its expression in nature. Darwin clearly recognized this – he was way ahead of his time. Which is why it is a shame he chose to exclude his beloved barnacles from his musings on the manifestation of sex. They might have prevented him from presenting sex in such a dichotomous and deterministic fashion. Today barnacles, and creatures like them, are at the forefront of teaching us how sex is no static binary, but a fluid phenomenon, with fuzzy borders that can bend to evolution's whim with astonishing speed.

The animal kingdom displays a smorgasbord of sexual flavours, with every form you can possibly imagine represented, and many you probably wouldn't conjure even in your wildest fantasies. One of the foremost scientists to broadcast the value in studying all this glorious variety is the theoretical ecologist Dr Joan Roughgarden, whose book *Evolution's Rainbow*, published in 2004, was the first to document and decode much of this diversity.

Over a lunch of tuna sandwiches and homemade pickles prepared by her husband Rick at their home in Hawaii, Roughgarden regaled me with tales of hermaphroditic nematode worms that reproduce by 'selfing' (yes, that is a thing), the homosexual societies of macho gay bighorn sheep and intersex bears that give birth through the tip of their 'penis'.

Roughgarden, now in her seventies and retired, has a personal interest in non-binary creatures. She started her academic career as Jonathan Roughgarden and transitioned while at Stanford in the late 1990s.* Joan Roughgarden has been a very noisy critic of Darwin's sexual selection theory. Its legacy, she claims, has forced generations of biologists to try to cram the immense sexual variation in nature – the rainbows, as she calls them – into overly orthodox binary boxes.

'The biggest error of biology today is uncritically assuming that the gamete size binary implies a corresponding binary in body type, behaviour and life history,' she states at the start of her book.

This has dangerous implications for science and society; 'suppressing the full story of gender and sexuality denies diverse people their right to feel at one with nature', she argues. 'The true story of nature is profoundly empowering for peoples of minority gender expressions and sexualities.'

Roughgarden's book was one of the first to point out that sexual differentiation is a complex process that involves the interaction of many genes and hormones. Subtle changes in their expression, which can be influenced by the environment as well as other genes, will influence an animal's sexual trajectory and lead to numerous equally viable and stable outcomes. As we discovered in chapter one, this inherent plasticity allows for much more variation in the expression of traits associated with sex and more overlap between the sexes than is commonly recognized, which fuels evolution. Sex is not all black or white, and labelling grey areas as anomalies – or worse, pathologies – means we fail to appreciate the natural function of diversity.

Roughgarden's iconoclastic thinking challenged the heteronormative straitjacket of Darwin's sexual selection theory, which considers the only role of sex to be procreation. Through this lens homosexual activity is denigrated to an inconvenient 'error' and subsequently ignored. Building on the work of the Canadian biologist Bruce Bagemihl, whose landmark modern bestiary catalogued

---

* 'I came out to Condoleezza Rice, who just happened to be my provost at the time,' she told me. 'I'm the only leftie in America who has anything nice to say about her, but she was pretty decent to me.'

homosexual activity in over three hundred species of vertebrate, Roughgarden extolled the role of same-sex activity in fostering cooperation amongst animal societies. We've seen this social glue at work with the bonobos, where sexual pleasure regulates social tension and promotes female–female alliances. Roughgarden includes numerous other species, from a wide range of taxa, where homosexual activity has, in her opinion, evolved as 'a social-inclusionary trait'.* These include a long-term study in the Netherlands where European oystercatchers, the monochrome marine birds common to many coastlines, have been documented nesting in 'threesomes' featuring two females and a male. These family units can be either aggressive or harmonious, depending on whether the females in the group are having sex with one another or not.

Such breeding strategies are generally branded as 'alternative' for betraying the nuclear family ideal widely assumed to be nature's norm. Roughgarden queries this pejorative label and argues how countless animal families frequently fail to conform to this 'Noah's Ark view'. In some species males and females have multiple sexual forms and identities, which she argues should be

---

* There are likely many adaptive outcomes for same-sex activity. Just as heterosexual sex is not always motivated by procreation alone (as demonstrated by Hrdy's promiscuous female langurs using sex as protection against infanticide) so same-sex activity can serve many socio-sexual purposes that fall outside of Darwin's original theory. Amongst male dolphins sex is considered a means of dominance negotiation. Amongst female macaques sexual activity encourages reconciliation after conflict. For male olive baboons the mutual handling of each other's vulnerable bits is considered an expression of mutual trust which stimulates cooperation, much like gang members 'swearing an oath'. There are many other theories as to the proximate cause of same-sex behaviour amongst animals that include genetics, development and life history. The most recent compendium of research into the widespread occurrence of same-sex activity amongst animals concluded that there is no one cause or effect and such behaviour can only be understood as a diverse collection of multicausal phenomena. From an evolutionary standpoint, a 2019 study postulated that at the dawn of sex the ancestral sexual drive may have been indiscriminate and directed towards all sexes. This subtle shift in perspective removes any kind of 'evolutionary conundrum' associated with the existence of same-sex sexual behaviour. Across the animal kingdom same-sex sexual activity is thus just as normal and expected as heterosexual behaviour is.

considered as different genders. Most biologists have shied away from applying this human cultural and psychological construct to the animal kingdom. In Roughgarden's book, however, gender in non-human animals has a different characterization, which she defines as 'the appearance, behaviour, and life history of a sexed body'. *Evolution's Rainbow* cites many species – from salmon to sparrows – that Roughgarden considers to have three, four or five separate genders; that is, animals that belong to the same biological sex but that have distinct appearances and sexual behaviours.

Take the bluehead wrasse (*Thalassoma bifasciatum*) – a colourful hermaphroditic fish that Roughgarden considers to have three genders. One gender is born male and remains male for life, another is born female and remains female. The third starts life as a female but later changes into a male. These sex-changed males are significantly larger than those that started life as male. They guard territories and females aggressively. Small unchanged males succeed at sex by forming coalitions and through teamwork manage to win matings.

Different environments favour different male genders. Big sex-changed males living amongst the cover of seagrass struggle to guard females, and the small unchanged males get more action. Whereas those big sex-changed males that live amongst the clear water of the coral reef fare better at guarding and have more success. So evolution favours a mixture of the two.

Many academics welcomed Roughgarden's radical proposal about animal gender and how it stirred fresh debate around the variety of sexual expression and its non-linear relationship with sex. Some of her other ideas have been less palatable. Roughgarden's accusation that sexual selection 'is an elite male heterosexual narrative projected on to animals' is harsh, but fair enough. Her insistence that Darwin was fundamentally wrong, however, and that sexual selection is 'false' has been less well received.

Roughgarden called for Darwin's 'conceptual rot' to be ripped up and replaced with a model that embraces nature's sexual smorgasbord, namely her theory of social selection. This revolutionary concept faltered at the first hurdle from being named after the established, and quietly scholarly, theory of social selection by Mary Jane West-Eberhard that we met in chapter seven. This caused unnecessary confusion as its proposal was quite different. Roughgarden placed group-level cooperation, rather than Darwinian competition, at the heart of sex – an unpopular idea that prompted no less than forty biologists to write to the journal *Science* defending Darwin's fundamental principles of within-sex competition and mate choice as drivers of sexual selection, if not his politics.

'I'm an academic terrorist,' she joked to me. 'In the UK, Darwin's not just a scientist, he's a national hero. It's part of British identity to praise Darwin's work. Which has led to a very conservative strain in British evolutionary biology.'

Roughgarden's ideas may not provide all the right answers; she's the first to admit that. What she has done is stimulate novel discourse around the entrenched archetypes of long-dead white men who viewed the world through a restrictive cultural lens. This can only be a good thing. Science thrives on debate and most evolutionary biologists would agree that sexual selection theory is in a state of metamorphosis. What conceptual butterfly finally emerges from this ongoing upheaval is the cause of vigorous debate. Roughgarden's instinct that studying, rather than ignoring, evolution's rainbows can offer valuable insights into sex and evolution was both timely and without contest. Her work has inspired others to embrace a queer theoretical perspective and reconsider archaic assumptions about the linear relationship between sex, sexuality and sexual expression in the animal kingdom. This is especially evident when you take a dive into the ocean and study the fish that live around coral reefs, which are as colourful sexually as they are visually.

If you go snorkelling on a reef, chances are a quarter of the fish you'll see are serial sex changers. The bluehead wrasse is just one of many rainbow-coloured reef fish that undergo a natural sex change during their adult life. Known as sequential hermaphrodites, these fish begin life as one sex and are triggered to switch camps by a social stimulus, such as the loss of a dominant individual or the relative availability of the opposite sex.

For some, like the flamboyant parrotfish, the change is permanent – once they've made the switch they're stuck as that sex until they die. Others have a more flexible sexuality and can switch back and forth throughout their life. This is a handy trick if, like the coral goby, you live in a crevice and are not that keen on venturing out much for fear of being eaten. If another coral goby comes along you can always reproduce, whatever its sex, by simply switching gonads to complement your new mate.

Some of these sex changers can flip with astonishing frequency. The chalk bass (*Serranus tortugarum*), a neon-blue Caribbean fish that's about the size of your thumb, has been known to switch sex up to twenty times a day. Chalk bass don't do this in order to play the field; quite the opposite – switching sex is their recipe for relationship success. Chalk bass are known to display unusual levels of sexual fidelity and are considered more or less monogamous. Their sex-change habits are a coordinated response with their long-term partner. Researchers believe that taking turns laying eggs, which are bigger and more energy-consuming to produce than sperm, keeps the reproductive investment fair. Each fish fertilizes as many eggs as it produces. Proving that even with fish you get what you give in a relationship.

Most sex changers are protogynous – meaning they start life as female and later become male. There are a handful of protandrous species, however, that do the reverse. The clownfish, also known as the anemonefish (from the subfamily *Amphiprioninae*), is one of the few that starts out male, and it's providing researchers with a unique opportunity to study the mechanisms behind what makes a female.

'The anemonefish is how we can study active feminization of the brain,' Justin Rhodes told me as he gave me a tour of his lab via Skype. 'We don't know anything about that and we should!'

The traditional model for sex differentiation, the Organizational Concept that we met in chapter one, considers brain feminization a passive process that simply happens by default in the absence of gonadal androgens. As a result, for the last seventy years science has been searching for brain dimorphisms that support the idea that, thanks to the masculinizing strength of testosterone in early development, male brains are from Mars and female brains from Venus. But with little luck. The anemonefish brain, which starts out male but becomes female, provides an opportunity to catalogue any neural changes.

Rhodes, a professor of psychology at the University of Illinois, is an engaging character, on the edge of eccentric, with an irrepressible enthusiasm for his anemonefish. I interviewed him online, and he whisked me around endless walls of aquaria filled to the gills with hundreds of orange-and-white striped fish, all apparently eyeballing me quizzically down the internet.

'This is one of my favourite females – look at how aggressive she is,' Rhodes said as he thrust his hand into a tank containing two small fish hovering around an upturned flowerpot containing a clutch of tiny eggs. The unhatched fry's silver eyeballs glinted through their cases.

'They're a very visual species. She can tell we're here now, and she's a little pissed off,' Rhodes added gleefully as his fingers were rammed by the plucky female, who I felt was more likely to be concerned by the scientist's big hairy hand than my hovering virtual head.

It was all quite surreal. Especially since the anemonefish have such a familiar face. These tiny two-toned fish became internationally famous as the stars of *Finding Nemo*. The smash-hit animation, released by Disney, featured the heart-warming tale of a young anemonefish (Nemo) that had lost his mum to a barracuda and ended up going on an incredible adventure before reuniting with his dad (Marlin).

Needless to say the movie takes more than a pinch of artistic licence with the real lives of anemonefish. These monogamous reef dwellers set up home together in an anemone, whose stinging tentacles offer the couple, and their eggs, protection. The belligerent female is the boss in the relationship – it's her job to defend the territory while the male cares for the eggs. The fish live surprisingly long lives – up to thirty years – in the same anemone, often with a bunch of juvenile males in attendance. If the female is removed, say nabbed by a barracuda, it will trigger Mr Anemonefish to transform into the new dominant female, and one of the juvenile males to mature into her mate.

A biologically accurate version of this hit movie would therefore have seen Nemo's father, Marlin, transition into a female, and then start having sex with her son, which might have made for a less popular family film amongst Disney's die-hard conservative audiences.

Rhodes' research has shown that the male-to-female sex change starts first in the brain and it's only later, after months or even years, that the fish's gonads catch up and become fully female.

'I was shocked,' Rhodes said. 'I had this one fish changing to a female. We kept taking blood every month, and we were like, this *can't* be right. They're showing male hormones, and then we'd do it again. I have thirty blood samples from this one fish and even after almost three years it still had male-like gonads.'

Rhodes introduced me to a fish just a few months into its change. The fish looked straight at me as Rhodes launched into graphic detail about the state of its gonads and brain. The trigger for switching sex starts in the brain as the fish perceives the change in its environment. This somehow activates a specific area of the brain, the pre-optic area, to increase in size. This region controls the gonads in all vertebrates. Females have around two to ten times more neurons in this region than males in order to manage the complicated production and release of eggs. Rhodes has discovered it takes up to six months for the pre-optic area to reach its full female size in anemonefish. In the meantime the testes might start shrivelling but are still producing androgens. They will do so until they have degenerated and been replaced by ripe egg-filled ovaries, many months later.

The transitioning fish thus had a female brain but male gonads. Which could be considered confusing for the fish, but Rhodes is certain that, if asked, the fish would say it's female.

'That's the great thing about these fish: they will tell you,' he told me. All he has to do is put the fish in a tank with another female. In the wild, these highly territorial fish refuse to share an anemone with another female, and they are guaranteed to fight, often to the death.

Rhodes sent me a video of two females engaged in battle, alerting me to listen out for them 'yelling at each other'. I had no idea fish could vocalize, let alone quarrel so vociferously, but Rhodes was right. The video showed two fish making an unexpected and quite aggressive sound, like popping popcorn, before getting more physical with one another. It was unmistakable behaviour. Quite different to a female encountering another male, which, as the submissive sex, does not instigate a deadly territorial war.

The simple experiment shows that these transitional anemone-fish both behave like females and are recognized as females by other fish, even though they have testes. It's a very clear demonstration that brain sex, and thus all sexual behaviour, and gonadal sex can be uncoupled. This begs the question, should the fish's sex be assigned by its gonads or its brain?

'If you said that fish was a male that would be wrong,' Rhodes explained. 'Whether it's a full-blown female, who cares? It will be eventually. The ovaries don't matter. They are female animals in terms of their behaviour and how they're recognized by other fish.'

Science, as it stands, would not agree. The fish's testes, albeit in a withering state, still trump its blossoming female brain and render it male in the eyes of many biologists. This exposes the flaw in assuming a linear relationship between gonadal sex, sexual identity, sexuality and sexed behaviour, even in a fish.

'The anemonefish question the way that we assign sex,' Rhodes added. 'The message of these fish is you should not define sex by the gonads.'

The juvenile 'males' pose an even greater classification quandary to Rhodes. Their immature gonads are officially 'ovotestes' with the potential to develop into male or female reproductive organs. They have some testicular tissue, but it's not actively producing sperm, and they also have ovarian tissue with undeveloped eggs. 'I don't think we can easily assign a sex to these non-breeders,' said Rhodes. 'They are neither male nor female.'

There are around five hundred known species of hermaphrodite fish. Likely many more. It's a nifty strategy to maximize your reproductive chances in the vastness of the ocean. Most are sequential hermaphrodites but some are simultaneously male and female. A select few, like the mangrove rivulus (*Kryptolebias marmoratus*), can even self-fertilize.

Hermaphrodites are widely distributed across all taxonomic groups of fish, including the most ancient. The terrifying-looking hagfish (of the class Myxini) is a muscular snake-like fish which, once seen, will haunt your nightmares. Also known as the slime eel, it has no scales, no spine and no jaw, but it does have whorls of sharp teeth for mincing the flesh of any decaying critters it sniffs out on the ocean floor using its solitary nostril.

As a lonely bottom feeder ploughing its slimy trade in the bleakness of the ocean depths it can be hard to find love, and several species of hagfish are known to spread their sexual options by sporting both an ovary and a testis. The hagfish hasn't changed for some 300 million years and is considered a primitive ancestor of modern fish, suggesting to Rhodes that hermaphroditism could be the ancestral state of not just fish, but all vertebrates.

'All these fish cast doubt on our binary system,' Rhodes explained.

The gonads of hermaphroditic fish may be hard to sex, but Rhodes has discovered their brains aren't radically different either. In his anemonefish Rhodes noted variance in the pre-optic area, and assumes there must be further morphological disparities associated with their behaviour – males do 80 per cent of the egg caring, while females do most of the fighting. But these roles are not rigidly enforced. Males will chase away the occasional intruder and females will do some caring for eggs, so there will be a corresponding over-lap in brain development.

'I don't want to overemphasize these differences,' Rhodes told me. 'It's not like males have a completely different brain than females. Most of the brain is very, very similar.'

This will come as no surprise to David Crews. The former profes-sor of zoology and psychology at the University of Texas whom we met in chapter one has devoted his career to studying sex and sexu-ality in animals and is considered one of the world's leading experts.

'There's a fundamental bisexuality* in every vertebrate organism,' Crews told me over the phone during one of our many long conver-sations during lockdown. Hermaphrodites like the anemonefish are perfect illustrations of this inherent bisexuality. Since they can adopt both male and female roles, and switch throughout adult life, they must have a bi-potential brain in terms of organization and function. Crews told me about an experiment by neurobiologist Leo Demski which neatly demonstrated this in the sea bass, a simultaneous her-maphrodite. The same fish could be encouraged to release sperm by stimulating one brain area, whereas stimulation of another brain area caused egg release.

Like Justin Rhodes, Crews has also observed how an animal's sexual behaviour is unconnected to its gonadal sex. He first docu-mented this while studying the whiptail lizards we met in the last chapter. This unisexual species is all female and reproduces by cloning, but alternate male and female copulatory behaviours sug-gest their brains are wired for both.

'We need to get away from the binary nature of sex assignment,' Crews ventured. 'There's a continuum, with males at one end and females at the other, and variability is continuous between those two types.'

This view is echoed by Mary Jane West-Eberhard whose

---

* When Crews talks about bisexuality he is not referring simply to sexual orientation, he means that all sexes have the potential to develop male or female sexual characteristics, be they behavioural or morphological.

encyclopaedic tome on developmental plasticity recognizes a spectrum of continuous variation in sexual characteristics with many kinds of 'intersex' between the masculine and feminine extremes. Biology is nevertheless grappling with how to define animals like the transitional anemonefish we've just met, Darwin's barnacles or the frogs with mixed gonads and behaviours from chapter one. Clearly they cannot be crammed into one of two binary boxes. On the basis of their gonads, most would be defined as hermaphrodite or intersex – a definition which not only destroys the dichotomous definition of sex, but is itself overly simplistic. Even this third category doesn't help when it comes to deciding where one draws the line between female and intersex and male, which is inevitably both subjective and arbitrary. The mole we met in chapter one, for example, has been variously described by different scientific papers over the years as 'female', 'sex-reversed', 'hermaphrodite', 'intersex' or even a mixture of several terms. Thus rendering the mole's sex a confusion for science, if not for the animal.

Modern researchers have criticized scientific language for failing to evolve fast enough to encompass the spectrum of sexual systems and their expression. 'It is important to remember that the diversity found in nature will seldom be captured by our terminology,' opined the author of a recent linguistically challenged compendium of sexual systems. New terms like 'quantitative gender' and 'conditional sex expression' were cited as attempts to bridge the language gap, but they're not exactly catchy.

Whether the issue is semantic or philosophical the fact remains that mainstream biology is proving slow to evolve beyond basic binary definitions of sex and recognize the facts of biology, which is somewhat ironic.

'I've come to the conclusion that the human brain likes black-and-white examples. It likes things to be one thing or the other, but that's problematic when it comes to sex,' Crews offered as his explanation for this paradox.

It is Crews' belief that viewing the animal kingdom through binary goggles has forced scientists like Darwin, and many who followed in his footsteps, to focus on the differences between the sexes, when studying the parallels could be more revealing.

'People forget that the majority of traits in males and females are similar. We all have brains, we all have hearts, we all have our bodies. There are more similarities between the sexes than there are differences.'

Crews illustrated this profound and inspiring point, like all good scientists would, with a rather boring graph. The overlap between two humps of linear data (one for male and one for female) demonstrated how individual variation within a sex is greater than the average variation between the sexes. Biology often ignores individual variation and irons out any extremes to consider only what is typical for each sex. So on paper the sexes appear to be completely different, but this is just a statistical phenomenon. The truth is that males and females are more alike than they are different.

'We all come from a single fertilized cell. Therefore we have to have all of the elements to create both sexes,' Crews added. 'I feel that if we can study the similarities in greater depth we might discover that, in fact, everything about the individual is bisexual.'

The anemonefish would certainly agree. But so might some of the more binary creatures we've met in the pages of this book. The synchronized courtship and sexual behaviour of those all-female albatross or gecko couples and the neural switch for 'maternal instinct' demonstrated in the brains of male and female frogs and mice also suggest Crews is correct.

A recent study into nematode worms, published in 2020, offers a further body blow to the binary dogma. These microscopic roundworms are a model organism much-loved by neuroscientists seeking blueprints for the control of behaviour in more complex creatures, including humans. Researchers at the University of Rochester Medical Center have isolated a genetic switch in the worms' brain cells, which allows them to toggle between sex-specific traits according to the demands of their environment. Male nematodes are wired to seek out sex, whereas the 'females' (which are also hermaphrodites) are focused on sniffing out food. But if the males get too hungry they make a radical switch to behave like the opposite sex. Such plasticity blurs the line between the sexes and challenges the idea of sex as a fixed property. 'These findings indicate that, at the molecular level, sex isn't binary or static, but rather dynamic and flexible,' proclaimed Douglas Portman, lead neuroscientist on the study.

Throughout this book we've met dozens of females that defy Darwin's rigid binary stereotypes. Female moles swilling with testosterone sporting bulging male gonads and no apparent vagina; aggressively dominant female hyenas with swinging pseudo-penises; female bearded dragons that are chromosomally male yet more fecund than their genetically female counterparts; post-menopausal orcas with socially and sexually purposeful lives; warring naked mole rat sisters tearing each other to death over supreme status; diminutive bonobo females that dominate males through ecstatic same-sex frottage; promiscuous female langurs whose profligate sex lives are a manifestation of extreme maternal care; and transitional female anemonefish still waiting to develop their ovaries.

These females teach us that sex is no crystal ball. It is neither static nor deterministic but a dynamic and flexible trait, just like any other, that's shaped by the peculiar interaction of shared genes with the environment, further sculpted by an animal's developmental and life histories plus a sprinkling of chance. Rather than thinking of the sexes as wholly different biological entities, we should consider them members of the same species, with fluid, complementary differences in certain biological and physiological processes associated with reproduction, but otherwise much the same. The time has come to ditch damaging, and frankly deluded, binary expectations because, in nature, the female experience exists on a genderless continuum: it is variable, highly plastic, and refuses to conform to archaic classifications. Our appreciation of this fact can only enrich our understanding of the natural world and empathy for one another as humans. Maintaining a dogged belief in antiquated sex differences only serves to fuel unrealistic expectations of women *and* men, foster poor inter-sexual relations and promote sexual inequality.

CONCLUSION

A *natural world without prejudice*

'Objective knowledge' is an oxymoron.
Patricia Gowaty, *Feminism and Evolutionary Biology*

When I had the first inkling to write a book about how female animals had been misrepresented by science, I had no idea the story was so big or that my subject was so vulnerable to cultural pollution. I was under the impression that science was, well, scientific. That is, rational, evidence-based, empirically deduced and uncontaminated knowledge. That much of what I had been taught as gospel at university – the very foundations of evolutionary biology – had been distorted by prejudice was a shocking revelation. One which has forced me to confront my own biases, and made me wonder if we can ever escape the shackles of personal perception and view the animal world through truly impartial eyes.

I am not the first person to ask this question. Even the Victorian academic establishment was aware that scientific knowledge is socially constructed. Over three decades before Darwin published *The Descent of Man*, the visionary polymath William Whewell, who amongst other notable achievements gifted the word 'scientist' to the English language, cautioned against this in one of his many philosophical musings on science.

'There is a mask of theory over the whole face of nature . . .

Most of us are unconscious of our perpetual habit of reading the language of the external world and translating as we read.'

What makes this mask so hard to remove is its invisible nature. We are all culturally conditioned to interpret the world within a framework of understanding that's both deeply ingrained and highly personal. Stepping outside this safety net of certainty requires first recognizing it exists. Then being brave enough to admit you are out of your depth, but to keep on swimming forward regardless.

It has taken the biological sciences a long time to face up to these failings. Feminism has played a key part. The initial wave started to build during the tail end of Darwin's career and his sexual selection theory came under attack from those first pioneers of equal rights. Four years after the publication of *The Descent of Man* the American minister and self-taught scientist Antoinette Brown Blackwell published *The Sexes Throughout Nature*, in which she argued that Darwin had misinterpreted evolution by giving 'undue prominence to such as have evolved in the male line'. The more complex or advanced the organism, she suggested, the greater the division of labour between the sexes. For every special character males evolved, females evolved complementary ones. The net effect being 'organic equilibrium in physiological and psychological equivalence of the sexes'.

Blackwell was no lone voice. A handful of women intellectuals, all self-educated, read Darwin's work and recognized that the female of the species had been marginalized and misunderstood. But these early feminist voices were ignored by the scientific patriarchy. Science was the preserve of the 'rational sex' in the Victorian era.* As Sarah Blaffer Hrdy has wryly noted, the impact of these feminist forerunners on mainstream evolutionary biology can be summed up with one phrase: 'The road not taken.'

Thankfully, Hrdy's voice, along with the other twentieth-century

---

* Darwin's letters suggest that, although he was open to the concept of women being involved in science, his default position was that it was primarily a male domain. When Blackwell wrote to Darwin with a copy of another of her scientific books, *Studies in General Science*, she hid her sex by signing her forename using initials only. Darwin's response to her was addressed 'Dear Sir . . .'

feminist scientists, was finally heard, albeit after much shouting. These women benefited from an egalitarian education and the associated intellectual confidence that enabled them to take on the second wave of scientific sexism, peddled by notorious neo-Darwinist male evolutionary biologists and psychologists. Their groundbreaking work has helped forge a radical shift in our understanding of not just what it means to be female, but of evolutionary theory itself.

You have met some of these pioneering scientists in the pages of this book. There are, of course, many others, of all sexes and genders, that could have been included. Thanks to their fearless logic we've moved beyond a rigid deterministic view of sex to appreciate how developmental plasticity and behavioural variation fuels the evolution of females, just as much as males. And how the mechanics driving that evolution are a tangled mix of natural, sexual and social selection. In addition to male–male competition and female choice, it is clear that female–female competition over mates and resources, male choice, strategic female cooperation with *both* sexes, and sexually antagonistic co-evolution all have the power to shape mating success.

I don't really understand why the names of the most published and prolific of these revolutionary scholars – Hrdy as well as Patricia Gowaty, Jeanne Altmann and Mary Jane West-Eberhard, amongst others – are not more widely recognized for the scientific and cultural impact of their work. In my mind, they deserve to be as famous as their male counterparts – Robert Trivers, Richard Dawkins, Stephen Jay Gould et al. But for some reason, even though their ideas are now woven into modern evolutionary thinking, the female authors of these bold new perspectives are still relatively invisible.

On one of our many long phone calls, Gowaty, whose work has successfully exposed, and sought to replace, much of the bigotry that plagues evolutionary theory, expressed an almost tearful gratitude for my interest in her work, lamenting, 'I'll be famous after I am dead.' I hope this book helps give these innovative thinkers the recognition they deserve.

The mask of theory is indeed slipping, but there is still much work to be done. The roots of a century of chauvinism have embedded themselves in the very building blocks of evolutionary thought. Bateman's paradigm continues to prevail in scientific literature, with little or no reference to the published empirical criticism by Gowaty et al. The textbooks used to educate the next generation of evolutionary biologists are still weighted towards an outdated male perspective on sexual selection. A 2018 analysis found that images of males and females in evolution textbooks still reinforced stereotypical sex roles and 'did not reflect the shift happening in the scientific community'.

Bias lurks in language too. A recent study found that science authors still use active words to describe males, but passive words to describe females – males have 'adaptations' and females 'counter-adaptations'. In other words, males act, whereas females react. Stereotypical labels – 'caring' females and 'competitive' males – are still bandied about in academic literature as if they were incontrovertible facts with no apparent need for back-up citations to justify their use.

Research is still directed towards males, whether intentionally or not. The 'type specimens' that define a species, and form the great library of life in the world's natural history museums, are heavily biased towards males. In a world where the living kind is increasingly rare these stuffed, cured and pickled archetypes form the basis for research, and a lack of female representation projects further androcentric skew into the future of evolutionary biology, ecology and conservation. Patricia Brennan's work documenting vagina and clitoral diversity is a much-needed initiative, but the rest of the female needs cataloguing, alongside the males.

In laboratory studies of living animals, many researchers avoid using females. Our 'messy hormones' are seen to complicate matters and males are considered more pure. This insidious myth is pure bunkum. There is no strong evidence that the oestrus cycle creates any greater endocrine variation in oestrogen than is seen for testosterone in males. A female's hormones are no messier than a male's.

When females are used, they aren't necessarily representative either. The Harvard neuroscientist Catherine Dulac told me how the little white mice that form the basis of much laboratory research have been domesticated by chauvinism. In the wild female mice are just as aggressive as males – they attack unknown suitors and cannibalize babies. But their aggressive nature has been bred out over the years to create a model animal that represents 'what a female should be'. This fake female then forms the model basis of much lab-based behavioural and neurobiological research. Studies of wild animals, instead of laboratory-bred, may be more challenging in terms of funding and fieldwork, but as Lauren O'Connell (and her dart frog parents) or David Crews (and his unisexual lizard) will tell you, the effort is necessary in order to study an unpolluted system.

Even when model organisms aren't bred towards a patriarchal ideal, they can still uphold those norms. Model systems are meant to provide general results for a given aspect of biology, but the species chosen are often questionable. Their use is often trapped by history and convenience, rather than relevance. Fruit flies, for example, still dominate sexual selection research and account for almost a quarter of all the papers published on this subject. Yet they fell into that model role purely because their breeding cycle is compatible with the academic calendar. Thus helping scientists more than the science they practise. The fact that their sexual behaviour is not representative of other insects, let alone taxa, hasn't stopped their eccentric kinks being extrapolated as sexual role models for all animals, even us.

TRUTH LIES IN DIVERSITY
AND TRANSPARENCY

Marlene Zuk, professor of evolutionary biology at the University of Minnesota, has cautioned against the use of improper model systems and campaigned for diversity; both in terms of wild and captive specimens, but also the range of species used. She argues that if we were starting from first principles we would never have chosen fruit flies as the basis of sexual selection research, and that their quirks have helped fuel confirmation bias. She's cautioned how 'taxonomic chauvinism' is a very real thing and means that certain animal groups – namely insects and birds – that are either charismatic or convenient, have dominated sexual selection research, ironing out natural diversity. Which is particularly paradoxical for evolutionary biology, with its inherent interest in variation.

We also need diversity in the people producing science. The rules of evolutionary biology weren't just developed by men, but by white, upper-class men from Western post-industrial societies. A mixture of sexes, sexualities, genders, skin colours, classes, cultures, abilities and ages working together on research projects will help flush out biases of all kinds, be they sexist, geographical, heteronormative, racist or otherwise. We need to be better at attracting these voices, and encouraging them to stay. A recent study found that LGBTQ people in STEM are still statistically underrepresented, encounter non-supportive environments and leave at an alarming rate. And women, despite decades of feminism, are still having to fight for equal opportunities when it comes to promotion and research funding. The stale old propaganda concerning gender disparities in an innate aptitude for science still haunts female scientists, despite its startling lack of legitimacy.

The Victorian era was all about imposing order on the natural world by creating rules that reflected cultural norms. The latest generation of evolutionary biologists are learning how to embrace the chaos of individual flexibility, developmental plasticity and the limitless possibilities of the natural world. Many are not only thinking outside of those Victorian boxes, but figuring out how to banish the mask of theory for ever.

In the last five to ten years there has been a flurry of critical introspection. Meta-analyses and reviews of evolutionary biology papers have exposed how insidious bias hides in experiment design and practice, and provided recommendations for how to remove it.

An examination of almost three hundred evolutionary and ecology studies from 1970 to 2012 found that more than half failed to disclose full details of the experiments' results and statistics. Sample sizes were often too small to reflect a result that could be divorced from chance, yet outcomes were reported as significant. Available data suggests that questionable research practices are common enough to warrant concern. Hannah Fraser, an ecologist at the University of Melbourne, surveyed more than eight hundred ecologists and evolutionary biologists and found that many admitted to at least one instance of cherry picking statistically significant results, using the flexible stopping rule (when data is collected until the right result achieved), or of having changed hypotheses to fit their results. The worst offenders were mid-career and senior scientists, who really should know better.

These alarming results suggest a need for critical evaluation of existing work and the urgent replication of pivotal studies – as Gowaty did with Bateman's. Replication may be a cornerstone of science, but these studies can be hard to get funded and appear less sexy to a career scientist as they don't involve original work. Funding bodies and editors of scientific publications need to be on board to help overcome this stigma.

But despite all this I have hope.

Researching this book was a liberating experience. I no longer feel like a sad misfit. Females are not destined to be passive and coy, evolutionary afterthoughts just waiting to be dominated by males. Even when we are physically weaker, we can still be powerful. I shall never forget bobbing along in that tiny boat off the coast of Washington State and feeling so moved by the awesome presence of that socially minded and extraordinarily empathetic menopausal orca. She showed me how strength can come through wisdom and age – a story I found especially profound.

It can also come from communion with other females. The female bonobos' solidarity with one another was totally inspiring. I'm not suggesting we all start having sex with each other, but there are lessons to be learnt from their peaceful society. Both the bonobo and the orca matriarchs demonstrate how dominance and leadership are quite different things. One does not necessarily follow the other, and they can even co-exist.

But I think it was those transitional anemonefish that rocked my world most. They made me think about my sex in the most radical way, forcing me to question core presumptions about how it is defined. This was both uncomfortable and exhilarating. Discovering that biological sex is, in truth, a spectrum and that all sexes are the product of basically the same genes, the same hormones and the same brains, has been the greatest revelation of all. It's forced a shift in my perspective to recognize my own cultural biases and try to banish any lingering heteronormative presumptions about the relationship between sex, sexual identity, sexed behaviour and sexuality. A freedom of thought that's challenging to maintain, but leaves me empowered by the boundless possibilities of the female experience.

Along this intellectual journey I have spoken to many young scientists, of all sexes and genders, that also give me hope. Their generation seems so much more attuned to challenging long-standing binary presumptions about sex. They are vocal in articulating the diversity and transparency practices that could finally remove the mask of theory for good. This cannot happen soon enough. As the American author and academic Anne Fausto-Sterling remarked, 'Biology is politics by any other means.' Theories devised by old sexist white men best serve old sexist white male politicians. The fight for biological truth is crucial if we are to forge a more inclusive society that can work together to protect the future of our planet and all that live on it.

ACKNOWLEDGEMENTS

No book is easy, but *Bitch* has been a bitch to squeeze out: gigantic, daunting, painful, intellectually gruelling and personally challenging. Huge thanks to my editors, Susanna Wadeson and Thomas Kelleher, for their patience, support and belief in this beast of a book. Their teams on both sides of the Atlantic have done a great job – thanks to Bella Bosworth for her copyedit, Alison Barrow for generating buzz, Kate Samano and Melissa Veronesi for martialling the book into production. And especially Emma Berry for her insightful notes, which kicked the book up a gear or two. My agents Will Francis and Jo Sarsby both deserve a massive thank-you for their enduring support and faith in me.

The research for this book was monumental and I am indebted to the many brilliant brains who held my hand as we scaled that cerebral mountain. The groundwork was laid by some super-smart young academics: Anne Hilborn, Adriana Lowe and Mrinalini Erkenswick Watsa got me started with key ideas and insights. Then the diligent Jenny Easley became my right-hand researcher in the thick of things. Massive thanks have to go to my academic readers, Kelsey Lewis (Wittiq fellow in feminist biology at the University of Wisconsin–Madison) and Jakob Bro-Jørgensen (senior lecturer in evolution and animal behaviour at the University of Liverpool) whose thorough notes on each draft were completely invaluable.

My deepest gratitude goes to the scientists whose pioneering research informed this book. I am in awe of you all and so very grateful that you shared your precious time to answer my many questions and patiently explain your science, frequently more than once. I am humbled by your generosity, honesty and trust in equal measures. In particular, David Crews and Patty Gowaty, who indulged my questions more than most and chatted with me so often I feel we're now friends.

Special thanks to those scientists who were brave enough to let me join them in the field, at home or in the lab: Rebecca Lewis and Andrea Baden for the Madagascan lemur passport, Gail Patricelli and Eric Tymstra for the sage-grouse show, Amy Parish for introducing me to Loretta, Chris Faulkes for letting me fondle a mole rat, Patricia Brennan for the endless rubbery vaginas, Deborah Giles for teaching me how to catch whale scat, Molly Cummings for the gin and fierce fish, Amber Wright for the last-minute lizards, Lindsay Young for those amazing albatrosses, and Joan Roughgarden for the pickles and provocative chat. Last but not least, the amazing force that is Sarah Blaffer Hrdy for warmly welcoming me into such a select summit, baking me a special 'vulture' pie and guiding me with munificence from the very start of this mammoth journey. I am deeply indebted to you on so many levels.

The three years that it took me to write this book were tumultuous on a personal level. I lost my mother and had to cope with the uneasy solitude of pandemic life. I have my dog Kobi to thank for much-needed oxytocin, and my lockdown bubblers Ruth Illger and Drew Carr for my sanity. Special thanks to my sunrise-swim buddies – Luke Gottelier, Sarah Farinhia, Jemima Dury and Berry White – who braved icy seas to wash away the anxiety and replace it with belly laughs. And to Penny and Marcus Fergusson for a writer's bunker at Feltham's farm, complete with diet of deliciously stinky cheese. Many of my closest friends patiently listened while I chewed the facts into stories, or read drafts and offered advice: Sarah Rollason, Heather Leach, Bini Adams, Wendie Ottiewill, Rebecca Keane, Jess Search, Sara Chamberlain, Alexa Haywood and Charlotte Moore. Special thanks to Carole Cadwalladr for giving this Bitch a title. And Maxx Ginnane for challenging the binary dogma in a late-night conversation and forcing me to confront my own preconceptions. This book is about cultural bias and I am grateful to have such a smart and diverse bunch of bitches to help me keep my world view evolving. I love you all.

NOTES

Introduction

ix 'The female is exploited': Richard Dawkins, *The Selfish Gene* (Oxford University Press, 2nd edn, 1989; 1st edn, 1976), p. 146

ix 'It is possible to interpret all other differences', 'Female exploitation begins here': ibid., pp. 141–2

xi only incorporated by Darwin under duress: 'Survival of the Fittest', Darwin Correspondence Project (University of Cambridge), https://www.darwinproject.ac.uk/commentary/survival-fittest [accessed March 2021]

xii 'The males of almost all animals': Charles Darwin, *The Descent of Man, and Selection in Relation to Sex* (John Murray, 2nd edn, 1879; republished by Penguin Classics, 2004), pp. 256–7

xii 'weapons' or 'charms' specially evolved in order to take 'possession': Charles Darwin, *On the Origin of Species* (John Murray, 1859; republished by Mentor Books, 1958), p. 94

xii why this disparity existed: Darwin, *The Descent of Man*, p. 259

xii unpalatable to the scientific patriarchy: Helena Cronin, *The Ant and the Peacock* (Cambridge University Press, 1991)

xii 'comparatively passive': Darwin, *The Descent of Man*, p. 257

xii 'standing by as spectators': Darwin, *On the Origin of Species*, p. 94

xiii 'In those animals that have . . . two sexes': Aristotle, *The Complete Works of Aristotle*, ed. by Jonathan Barnes (Princeton University Press, 2014), p. 1132 xiv 'someone to take care of the house', 'nice soft wife', 'better than a dog': *The Autobiography of Charles Darwin*, ed. by N. Barlow (New York, 1969), pp. 232–3

xiv the prevailing chauvinism of the era: Evelleen Richards, 'Darwin and the Descent of Woman' in *The Wider Domain of Evolutionary Thought*, ed. by David Oldroyd and Ian Langham (D. Reidel Publishing Company, 1983)

xv males were considered to be more *evolved*: Zuleyma Tang-Martínez, 'Rethinking Bateman's Principles: Challenging Persistent Myths of Sexually Reluctant Females and Promiscuous Males', *Journal of Sex Research* (2016), pp. 1–28

xv 'Thus man has ultimately become superior': Darwin, *The Descent of Man*, pp. 629/631

xvi they went in search of the 'alpha male': John Marzluff and Russell Balda, *The Pinyon Jay: Behavioral Ecology of a Colonial and Cooperative Corvid* (T. and A. D. Poyser, 1992), p. 110

xvi 'There is little doubt that adult males': ibid., p. 113

xvii 'the avian equivalent of PMS': ibid., pp. 97–8

xvii 'the crowning of a new king': ibid., p. 114

xvii There is no conspiracy here, just blinkered science: Marcy F. Lawton, William R. Garstka and J. Craig Hanks, 'The Mask of Theory and the Face of Nature' in *Feminism and Evolutionary Biology*, ed. by Patricia Adair Gowaty (Chapman and Hall, 1997)

xvii view the animal kingdom from their standpoint: William G. Eberhard, 'Inadvertent Machismo?' in *Trends in Ecology & Evolution*, 5: 8 (1990), p. 263

xviii most dangerous thing about sexist bias: Hillevi Ganetz, 'Familiar Beasts: Nature, Culture and Gender in Wildlife Films on Television' in *Nordicom Review*, 25 (2004), pp. 197–214

xviii should stick to mothering: Anne Fausto-Sterling, Patricia Adair Gowaty and Marlene Zuk, 'Evolutionary Psychology and Darwinian Feminism' in *Feminist Studies*, 23: 2 (1997), pp. 402–17

xviii 'not a single well-known feminist': ibid.

Chapter One: The anarchy of sex

2 consume over half her body weight in worms a day: 'Species – Mole', Mammal Society, https://www.mammal.org.uk/species-hub/full-species-hub/discover-mammals/species-mole/ [accessed 5 May 2021]

2 modified form of haemoglobin: Kevin L. Campbell, Jay F. Storz, Anthony V. Signore, Hideaki Moriyama, Kenneth C. Catania, Alexander P. Payson, Joseph Bonaventura, Jörg Stetefeld and Roy E. Weber, 'Molecular Basis of a Novel Adaptation to Hypoxic-hypercapnia in a Strictly Fossorial Mole' in *BMC Evolutionary Biology*, 10: 214 (2010)

2 she sports an extra 'thumb': Christian Mitgutsch, Michael K. Richardson, Rafael Jiménez, José E. Martin, Peter Kondrashov, Merijn A. G. de Bakker and Marcelo R. Sánchez-Villagra, 'Circumventing the Polydactyly "Constraint": The Mole's "Thumb"' in *Biology Letters*, 8: 1 (23 Feb. 2012)

2 the testicular tissue expands until it is actually larger than the ovarian: Jennifer A. Marshall Graves, 'Fierce Female Moles Have Male-like Hormones and Genitals. We Now Know How This Happens', The Conversation, 12 Nov. 2020, https://theconversation.com/fierce-female-moles-have-male-like-hormones-and-genitals-we-now-know-how-this-happens-149174

2 described as a 'phallus' or 'penile clitoris': Adriane Watkins Sinclair, Stephen E. Glickman, Laurence Baskin and Gerald R. Cunha, 'Anatomy of Mole External Genitalia: Setting the Record Straight' in *The Anatomical Record* (Hoboken), 299: 3 (March 2016), pp. 385–99

3 considered the preserve of humans: David Crews, 'The Problem with Gender' in *Psychobiology*, 16: 4 (1988), pp. 321–34

3 This basic gametal dichotomy: Joan Roughgarden, *Evolution's Rainbow* (University of California Press, 2004), p. 23

3n evolved on two independent occasions: Kazunori Yoshizawa, Rodrigo L. Ferreira, Izumi Yao, Charles Lienhard and Yoshitaka

Kamimura, 'Independent Origins of Female Penis and its Coevolution with Male Vagina in Cave Insects (Psocodea: Prionoglarididae)' in *Biology Letters*, 14: 11 (Nov. 2018)

4 exudes yellow liquid: Clare E. Hawkins, John F. Dallas, Paul A. Fowler, Rosie Woodroffe and Paul A. Racey, 'Transient Masculinization in the Fossa, *Cryptoprocta ferox* (Carnivora, Viverridae)' in *Biology of Reproduction*, 66: 3 (March 2002), pp. 610–15

4 scientific paper on fossa genitalia: ibid.

5 inspect one another's sexual tumescence during 'greeting ceremonies': Christine M. Drea, 'Endocrine Mediators of Masculinization in Female Mammals' in *Current Directions in Psychological Science*, 18: 4 (2009), pp. 221–6

5 'palpation of the scrotum': Paul A. Racey and Jennifer Skinner, 'Endocrine Aspects of Sexual Mimicry in Spotted Hyenas *Crocuta crocuta*' in *Journal of Zoology*, 187: 3 (March 1979), p. 317

6 Females are considered dominant in most situations: Alan Conley, Ned J. Place, Erin L. Legacki, Geoff L. Hammond, Gerald R. Cunha, Christine M. Drea, Mary L. Weldele and Steve E. Glickman, 'Spotted Hyaenas and the Sexual Spectrum: Reproductive Endocrinology and Development' in *Journal of Endocrinology*, 247: 1 (Oct. 2020), pp. R27–R44

6n The most extreme mammalian case: Katherine Ralls, 'Mammals in which Females Are Larger than Males' in *The Quarterly Review of Biology*, 51 (1976), pp. 245–76

6n One female specimen: Richard Sears and John Calambokidis, 'COSEWIC Assessment and Update Status Report on the Blue Whale, *Balaenoptera musculus*' (Mingan Island Cetacean Study, 2002), p. 3

6n 'so perfect and complete is the union': Theodore W. Pietsch, *Oceanic Anglerfishes: Extraordinary Diversity in the Deep Sea* (University of California Press, 2009), p. 277

7 'a crystal of androgen could counteract the absence of testicles': Anne Fausto-Sterling, *Sexing the Body* (Basic Books, 2000), p. 202

8 This then programmed sexual differences in bodies and behaviour: Charles H. Phoenix, Robert W. Goy, Arnold A. Gerall

and William C. Young, 'Organizing Action of Prenatally Administered Testosterone Propionate on the Tissues Mediating Mating Behavior in the Female Guinea Pig' in *Endocrinology*, 65: 3 (1 Sept. 1959), pp. 369–82

8 'Becoming a male is a prolonged, uneasy and risky adventure': Fausto-Sterling, *Sexing the Body*, p. 202

9 Females basically 'just happened': ibid.

9 'indistinguishable from that of males': J. Thornton, 'Effects of Prenatal Androgens on Rhesus Monkeys: A Model System to Explore the Organizational Hypothesis in Primates' in *Hormones and Behavior*, 55: 5 (2009), pp. 633–45

10 transforming both her pudenda and her post-natal behaviour: Christine M. Drea, 'Endocrine Mediators of Masculinization in Female Mammals'

11 'a widespread understanding that no active genetic steps', 'a rather amazing situation': Dagmar Wilhelm, Stephen Palmer and Peter Koopman, 'Sex Determination and Gonadal Development in Mammals' in *Physiological Reviews*, 87: 1 (2007), pp. 1–28

12 he abandoned his cytology studies: Bill Bryson, *The Body* (Transworld Publishers, 2019)

13 matures at a more leisurely pace: Andrew H. Sinclair, Philippe Berta, Mark S. Palmer, J. Ross Hawkins, Beatrice L. Griffiths, Matthijs J. Smith, Jamie W. Foster, Anna-Maria Frischauf, Robin Lovell-Badge and Peter N. Goodfellow, 'A Gene from the Human Sex-determining Region Encodes a Protein with Homology to a Conserved DNA-binding Motif' in *Nature*, 346: 6281 (1990), pp. 240–4

13 'essence of maleness': Roughgarden, *Evolution's Rainbow*, p. 198

15 exploit the benefits of 'adaptive intersexuality': Francisca M. Real, Stefan A. Haas, Paolo Franchini, Peiwen Xiong, Oleg Simakov, Heiner Kuhl, Robert Schöpflin, David Heller, M-Hossein Moeinzadeh, Verena Heinrich, Thomas Krannich, Annkatrin Bressin, Michaela F. Hartman, Stefan A. Wudy and Dina K. N. Dechmann, Alicia Hurtado, Francisco J. Barrionuevo, Magdalena Schindler, Izabela Harabula, Marco Osterwalder, Michael Hiller, Lars Wittler, Axel Visel, Bernd Timmermann, Axel Meyer, Martin Vingron,

Rafael Jimémez, Stefan Mundlos and Darío G. Lupiáñez, 'The Mole Genome Reveals Regulatory Rearrangements Associated with Adaptive Intersexuality' in *Science*, 370: 6513 (Oct. 2020), pp. 208–14

15 This egg-laying mammal . . . specializes in being contrary: Frank Grützner, Willem Rens, Enkhjargal Tsend-Ayush, Nisrine El-Mogharbel, Patricia C. M. O'Brien, Russell C. Jones, Malcolm A. Ferguson-Smith and Jennifer A. Marshall Graves, 'In the Platypus a Meiotic Chain of Ten Sex Chromosomes Shares Genes with the Bird Z and Mammal X Chromosomes' in *Nature*, 432 (2004)

16 platypus has five pairs of sex chromosomes: Frédéric Veyrunes, Paul D. Waters, Pat Miethke, Willem Rens, Daniel McMillan, Amber E. Alsop, Frank Grützner, Janine E. Deakin, Camilla M. Whittington, Kyriena Schatzkamer, Colin L. Kremitzki, Tina Graves, Malcolm A. Ferguson-Smith, Wes Warren and Jennifer A. Marshall Graves, 'Bird-like Sex Chromosomes of Platypus Imply Recent Origin of Mammal Sex Chromosomes' in *Genome Research*, 18: 6 (June 2008) pp. 965–73

16 before the human Y chromosome disappeared completely: Jennifer A. Marshall Graves, 'Sex Chromosome Specialization and Degeneration in Mammals' in *Cell* (2006), pp. 901–14

17 triggered by an entirely different . . . sex-determining gene: Asato Kuroiwa, Yasuko Ishiguchi, Fumio Yamada, Abe Shintaro and Yoichi Matsuda, 'The Process of a Y-loss Event in an XO/XO Mammal, the Ryukyu Spiny Rat' in *Chromosoma*, 119 (2010), pp. 519–26; E. Mulugeta, E. Wassenaar, E. Sleddens-Linkels, W. F. J. van IJcken, E. Heard, J. A. Grootegoed, W. Just, J. Gribnau and W. M. Baarends, 'Genomes of Ellobius Species Provide Insight into the Evolutionary Dynamics of Mammalian Sex Chromosomes' in *Genome Research*, 26: 9 (Sept. 2016), pp. 1202–10

17 an entirely new master switch gene: N. O. Bianchi, '*Akodon* Sex Reversed Females: The Never Ending Story' in *Cytogenetic and Genome Research*, 96 (2002), pp. 60–5

19 population density and social circumstance: Mary Jane West-
Eberhard, *Developmental Plasticity and Evolution* (Oxford
University Press, 2003), p. 121

20 three different 'sex races': Nicolas Rodrigues, Yvan Vuille, Jon
Loman and Nicolas Perrin, 'Sex-chromosome Differentiation
and "Sex Races" in the Common Frog (*Rana temporaria*)' in
Proceedings of the Royal Society B, 282: 1806 (May 2015)

21 found in herbicides like Atrazine: Max R. Lambert, Aaron
B. Stoler, Meredith S. Smylie, Rick A. Relyea, David K. Skelly,
'Interactive Effects of Road Salt and Leaf Litter on Wood Frog
Sex Ratios and Sexual Size Dimorphism' in *Canadian Journal of
Fisheries and Aquatic Sciences*, 74: 2 (2016), pp. 141–6

22 their behaviour and morphology are more masculine: Vivienne
Reiner, 'Sex in Dragons: A Complicated Affair' (University of
Sydney, 8 June 2016), https://www.sydney.edu.au/news-opinion/
news/2016/06/08/sex-in-dragons--a-complicated-affair.html
[accessed 10 April 2020]

22 a separate third sex: Hong Li, Clare E. Holleley, Melanie Elphick,
Arthur Georges and Richard Shine, 'The Behavioural Conse-
quences of Sex Reversal in Dragons' in *Proceedings of the Royal
Society B*, 283: 1832 (2016)

22 a powerful driver of evolutionary change: Clare E. Holleley,
Stephen D. Sarre, Denis O'Meally and Arthur Georges, 'Sex
Reversal in Reptiles: Reproductive Oddity or Powerful Driver of
Evolutionary Change?' in *Sexual Development* (2016)

22 Their 'male-like' brain appears to be driven by their inherent
genetic make-up: Li, Holleley, Elphick, Georges and Shine, 'The
Behavioural Consequences of Sex Reversal in Dragons'

23 having been eaten by Dr Schaef: Madge Thurlow Macklin, 'A
Description of Material from a Gynandromorph Fowl' in *Journal of Experimental Zoology*, 38: 3 (1923)

24 'It blew me away': Laura Wright, 'Unique Bird Sheds Light on
Sex Differences in the Brain', *Scientific American*, 25 March 2003

24 the sex chromosomes . . . must be playing a crucial role: Robert
J. Agate, William Grisham, Juli Wade, Suzanne Mann, John

Wingfield, Carolyn Schanen, Aarno Palotie and Arthur P. Arnold, 'Neural, Not Gonadal, Origin of Brain Sex Differences in a Gynandromorphic Finch' in *PNAS*, 100 (2003), pp. 4873–8

24 the genetic sexual identity of individual cells: M. Clinton, D. Zhao, S. Nandi and D. McBride, 'Evidence for Avian Cell Autonomous Sex Identity (CASI) and Implications for the Sex-determination Process?' in *Chromosome Research*, 20: 1 (Jan. 2012), pp. 177–90

27 the oestrogen receptor is the oldest transcription factor: J. W. Thornton, E. Need and D. Crews, 'Resurrecting the Ancestral Steroid Receptor: Ancient Origin of Estrogen Signaling' in *Science*, 301 (2003), pp. 1714–17

27 has reversed the sex of developing female lizards: David Crews, 'Temperature, Steroids and Sex Determination' in *Journal of Endocrinology*, 142 (1994), pp. 1–8

Chapter Two: The mysteries of mate choice

31 'queer antics': R. Bruce Horsfall, 'A Morning with the Sage-Grouse' in *Nature*, 20: 5 (1932), p. 205

32 'inconspicuous and passive': John W. Scott, 'Mating Behaviour of the Sage-Grouse' in *The Auk* (American Ornithological Society), 59: 4 (1942), p. 487

32 'It is curious that I remember well': Charles Darwin, letter to Asa Gray, 3 April 1860, Darwin Correspondence Project, https://www.darwinproject.ac.uk/letter/DCP-LETT-2743.xml

33 'It is shown by various facts ... the female ... exerts some choice': Charles Darwin, *The Descent of Man, and Selection in Relation to Sex* (John Murray, 2nd edn, 1879; republished by Penguin Classics, 2004), p. 257

33 'preference by the female of the more attractive males': Charles Darwin, *The Descent of Man, and Selection in Relation to Sex* (1871), vol. 1, p. 422

33 was without scientific precedent: G. F. Miller, 'How Mate Choice Shaped Human Nature: A Review of Sexual Selection and Human Evolution' in *Handbook of Evolutionary Pyschology: Ideas, Issues, and Applications*, ed. by C. Crawford and D. Krebs (1998), pp. 87–130

33 'a taste for the beautiful': Darwin, *The Descent of Man* (1871), p. 92

33 a human-like sense of aesthetics: Nicholas L. Ratterman and Adam G. Jones, 'Mate Choice and Sexual Selection: What Have We Learned Since Darwin?' in *PNAS*, 106: 1 (2009), pp. 1001–8

34 'surplus of strength, vitality, and growth-power': Alfred R. Wallace, *Darwinism* (Macmillan & Co., 1889), p. 293

34 'In rejecting . . . female choice': ibid., p. viii

34 Darwinist thinking in the twentieth century: Richard O. Prum, *The Evolution of Beauty: How Darwin's Forgotten Theory of Mate Choice Shapes the Animal World Around Us* (Anchor Books, 2017)

34 'the mad aunt in the evolutionary attic of Darwinian theory': ibid.

34 'the most dynamic areas': Thierry Hoquet (ed.), *Current Perspectives on Sexual Selection: What's Left After Darwin?* (Springer, 2015)

35 only 10–20 per cent of the males: A. Mackenzie, J. D. Reynolds, V. J. Brown and W. J. Sutherland, 'Variation in Male Mating Success on Leks' in *The American Naturalist*, 145: 4 (1995)

36 lose almost 7 per cent of their body mass: Jacob Höglundi, John Atle Kålås and Peder Fiske, 'The Costs of Secondary Sexual Characters in the Lekking Great Snipe (*Gallinago media*)' in *Behavioral Ecology and Sociobiology*, 30: 5 (1992), pp. 309–15

36 quietest directly in front of the displaying bird: Marc S. Dantzker, Grant B. Deane and Jack W. Bradbury, 'Directional Acoustic Radiation in the Strut Display of Male Sage Grouse *Centrocercus urophasianus*' in *Journal of Experimental Biology*, 202: 21 (1999), pp. 2893–909

40 emulate older (and perhaps wiser) female fishes' mate choice decisions: J. Amlacher and L. A. Dugatkin, 'Preference for Older Over Younger Models During Mate-choice Copying in Young Guppies' in *Ethology Ecology & Evolution*, 17: 2 (2005), pp. 161–9

41 females prefer the most nimble-minded male: Jason Keagy, Jean-François Savard and Gerald Borgia, 'Male Satin Bowerbird Problem-solving Ability Predicts Mating Success' in *Animal Behaviour*, 78: 4 (2009), pp. 809–17

42 'quite incredible . . . that a large majority of females': Alfred R. Wallace, 'Lessons from Nature, as Manifested in Mind and Matter' in *Academy*, 562 (1876)

42 by dressing up as her favourite food: Michael J. Ryan and A. Stanley Rand, 'The Sensory Basis of Sexual Selection for Complex Calls in the Túngara Frog, *Physalaemus pustulosus* (Sexual Selection for Sensory Exploitation)' in *Evolution*, 44 (1990), pp. 305–14

42 a partiality for the colour orange: F. Helen Rodd, Kimberly A. Hughes, Gregory F. Grether and Colette T. Baril, 'A Possible Non-sexual Origin of Mate Preference: Are Male Guppies Mimicking Fruit?' in *Proceedings of the Royal Society*, 269 (2002), pp. 475–81

42 suggesting their senses may be tuned for alert to this colour: Joah Robert Madden and Kate Tanner, 'Preferences for Coloured Bower Decorations Can Be Explained in a Nonsexual Context' in *Animal Behaviour*, 65: 6 (2003), pp. 1077–83

43 quest for 'hedonic pleasure': Michael J. Ryan, 'Darwin, Sexual Selection, and the Brain' in *PNAS*, 118: 8 (2021), pp. 1–8

43 'essentially mysterious': Gil Rosenthal, *Mate Choice* (Princeton University Press, 2017), p. 6

43 in the hope of getting their eggs fertilized before the pond party finished: Michael J. Ryan, 'Resolving the Problem of Sexual Beauty' in *A Most Interesting Problem*, ed. by Jeremy DeSilva (Princeton University Press, 2021)

43 sage grouse females . . . turn out to be surprisingly promiscuous: Krista L. Bird, Cameron L. Aldridge, Jennifer E. Carpenter, Cynthia A. Paszkowski, Mark S. Boyce and David W. Coltman, 'The Secret Sex Lives of Sage-grouse: Multiple Paternity and Intraspecific Nest Parasitism Revealed through Genetic Analysis' in *Behavioral Ecology*, 24: 1 (2013) pp. 29–38

Chapter Three: The monogamy myth

45 creeping away from her napping partner: P. Dee Boersma and Emily M. Davies, 'Why Lionesses Copulate with More than One Male' in *The American Naturalist*, 123: 5 (1984), pp. 594–611

45 mate up to one hundred times a day: Sarah Blaffer Hrdy, 'Empathy, Polyandry, and the Myth of the Coy Female' in *Feminist Approaches to Science*, ed. by Ruth Bleier (Pergamon, 1986), p. 123

46 'Excess copulations may not actually cost a female much': Richard Dawkins, *The Selfish Gene* (Oxford University Press, 2nd edn, 1989; 1st edn, 1976), p. 164

46 hens would routinely get down and dirty: Aristotle, *The History of Animals, books VI–X* (350 BC), trans. and ed. by D. M. Balme (Harvard University Press, 1991)

46 'In the most distinct classes of the animal kingdom': Charles Darwin, *The Descent of Man, and Selection in Relation to Sex* (John Murray, 2nd edn, 1879; republished by Penguin Classics, 2004), p. 272

47 'sedulously display their charms before the female': ibid., p. 256

47 'she may often be seen endeavouring': ibid, p. 257

47 foundations for 'active' masculinity and 'passive' femininity: Zuleyma Tang-Martínez, 'Rethinking Bateman's Principles: Challenging Persistent Myths of Sexually Reluctant Females and Promiscuous Males' in *Journal of Sex Research* (2016), pp. 1–28

48 Darwin's 'general law': Darwin, *The Descent of Man*, p. 257

48 'at a loss' to 'explain the sex difference': A. J. Bateman, 'Intrasexual Selection in *Drosophila*' in *Heredity*, 2 (1948), pp. 349–68

50 'an indiscriminating eagerness', 'discriminating passivity': ibid.

50 'Even in a derived monogamous species (e.g. man)': ibid.

51 'The Reluctant Female and the Ardent Male': Margo Wilson and Martin Daly, *Sex, Evolution and Behaviour* (Thompson/Duxbury Press, 1978)

51 'If you get caught fooling around': Erika Lorraine Milam, 'Science of the Sexy Beast' in *Groovy Science*, ed. by David Kaiser and Patrick McCray (University of Chicago Press, 2016), p. 292

51 Some even justify the worst human male behaviour: Craig Palmer and Randy Thornhill, *A Natural History of Rape* (MIT Press, 2000)

52 'recovered rapidly and without any apparent ill effects': Olin E. Bray, James J. Kennelly and Joseph L. Guarino, 'Fertility of Eggs Produced on Territories of Vasectomized Red-Winged Blackbirds' in *The Wilson Bulletin*, 87: 2 (1975), pp. 187–95

53 'Polyandry is unknown': David Lack, *Ecological Adaptations for Breeding in Birds* (Methuen, 1968)

54 a revolution in our understanding of female mating behaviour: Marlene Zuk, *Sexual Selections: What We Can and Can't Learn about Sex from Animals* (University of California Press, 2002), p. 64

54 the 'humble and homely' dunnock: Reverend F. O. Morris, *A History of Birds* (1856)

54 would have resulted in 'chaos in the parish': Nicholas B. Davies, *Dunnock Behaviour and Social Evolution* (Oxford University Press, 1992)

55 a single clutch of eggs can have many fathers: Marlene Zuk and Leigh Simmons, *Sexual Selection: A Very Short Introduction* (Oxford University Press, 2018), p. 29

55 over three quarters of the chicks: Tim Birkhead, *Promiscuity* (Faber, 2000), p. 40

55 her first taste of being ignored by the male scientific establishment: Zuleyma Tang-Martínez and T. Brandt Ryder, 'The Problem with Paradigms: Bateman's Worldview as a Case Study' in *Integrative and Comparative Biology*, 45: 5 (2005), pp. 821–30

57 female birds were 'suffering' forced extra-pair copulations: Tim Birkhead and J. D. Biggins, 'Reproductive Synchrony and Extra-pair Copulation in Birds' in *Ethology*, 74 (1986), pp. 320–34

57 'The more this oddly puritanical idea is examined, the stranger it appears': Susan M. Smith, 'Extra-pair Copulations in Black-capped Chickadees: The Role of the Female' in *Behaviour*, 107: 1/2 (1988), pp. 15–23

58 eventually published in 1997: Diane L. Neudorf, Bridget J. M. Stutchbury and Walter H. Piper, 'Covert Extraterritorial Behavior of Female Hooded Warblers' in *Behavioural Ecology*, 8: 6 (1997), pp. 595–600

58 'likely that females control the success of a copulation attempt': Marion Petrie and Bart Kempenaers, 'Extra-pair Paternity in Birds: Explaining Variation Between Species and Populations' in *Trends in Ecology & Evolution*, 13: 2 (1998), p. 52

58 a 'polyandry revolution': Zuk and Simmons, *Sexual Selection*, p. 32

59 less than 7 per cent of known species: Tang-Martínez and Brandt Ryder, 'The Problem with Paradigms'

59 'Generations of reproductive biologists': Birkhead, *Promiscuity*, p. ix

59 predicts that females have nothing to gain from 'excessive' matings: Hrdy, 'Empathy, Polyandry, and the Myth of the Coy Female'

59 'the myth of the coy female': ibid.

60 in the orbit of its wunderkind: Sarah Blaffer Hrdy, 'Myths, Monkeys and Motherhood' in *Leaders in Animal Behaviour*, ed. by Lee Drickamer and Donald Dewsbury (Cambridge University Press, 2010)

60 'invariably subordinate to all the adult males': Phyllis Jay, 'The Female Primate' in *Potential of Women* (1963), pp. 3–7

60 'a small part in the life of an adult female': ibid.

60 they were 'relatively identical': Hrdy, 'Empathy, Polyandry, and the Myth of the Coy Female'

60 'such wandering and such seemingly "wanton" behaviour': ibid.

61 their lustful advances are even refused: Sarah Blaffer Hrdy, *The Woman That Never Evolved* (Harvard University Press, 1981)

61 'In retrospect, one really does have to wonder': Hrdy, 'Empathy, Polyandry and the Myth of the Coy Female'

62 pronounced the human female orgasm to be 'unique among primates': Desmond Morris, *The Naked Ape* (Jonathan Cape, 1967)

62 'a fascination with her own genitals': Caroline Tutin, *Sexual Behaviour and Mating Patterns in a Community of Wild Chimpanzees* (University of Edinburgh, 1975)

62 'laughing softly as they do so': Alan F. Dixson, *Primate Sexuality: Comparative Studies of the Prosimians, Monkeys, Apes, and Humans* (Oxford University Press, 2012), p. 179

62 using their tails or 'soft surfaces' until they enter a 'trancelike' state: Phillip Hershkovitz, *Living New World Monkeys* (Chicago University Press, 1977), p. 769

62 concluded that females did indeed climax: Suzanne Chevalier-Skolnikoff, 'Male–Female, Female–Female, and Male–Male Sexual Behavior in the Stumptail Monkey, with Special Attention to the Female Orgasm' in *Archives of Sexual Behaviour*, 3 (1974), pp. 95–106

64 rhesus females do indeed have the ability to climax: Frances Burton, 'Sexual Climax in Female *Macaca Mulatta*' in *Proceedings of the Third International Congress of Primatologists* (1971), pp. 180–91

64 this orgasmic response is 'dysfunctional': Donald Symons, *The Evolution of Human Sexuality* (Oxford University Press, 1979), p. 86

64 'Are we to believe that the clitoris is nothing more than': Hrdy, *The Woman That Never Evolved*, p. 167

65 forces the bereaved mother into oestrus: Sarah Blaffer Hrdy, 'Male–Male Competition and Infanticide Among the Langurs of Abu Rajesthan' in *Folia Primatologica*, 22 (1974), pp. 19–58

65 'So, Sarah, put it another way': Claudia Glenn Dowling, 'Maternal Instincts: From Infidelity to Infanticide', *Discover*, 1 March 2003, https://www.discovermagazine.com/health/maternal-instincts-from-infidelity-to-infanticide

65 incorporated into mainstream academic thinking: Joseph Soltis, 'Do Primate Females Gain Nonprocreative Benefits by

Mating with Multiple Males? Theoretical and Empirical Considerations' in *Evolutionary Anthropology*, 11 (2002), pp. 187–97

65 murdering her cubs: Hrdy, 'Male–male Competition and Infanticide'

66 'flexible and opportunistic individuals': Sarah Blaffer Hrdy, 'The Optimal Number of Fathers: Evolution, Demography, and History in the Shaping of Female Mate Preferences' in *Annals of the New York Academy of Sciences* (2000), pp. 75–96

66 accurate in their paternity 'assessments': ibid.

67 'ultimate female reproductive strategy': Sarah Blaffer Hrdy, 'The Evolution of the Meaning of Sexual Intercourse', presented at Sapienza University of Rome, 19–21 Oct. 1992, sponsored by the Ford Foundation and the Italian Government

67 especially useful for our hominid ancestors: Hrdy, 'The Optimal Number of Fathers'

67 patriarchal social systems evolved in order to curb and confine it: Hrdy, 'The Evolution of the Meaning of Sexual Intercourse'

68 mirrors the different mating strategies of the females: Marlene Zuk, *Sexual Selections*, p. 80

68 a dainty pair of strawberries: G. J. Kenagy and Stephen C. Trombulak, 'Size and Function of Mammalian Testes in Relation to Body Size' in *Journal of Mammology*, 67: 1 (1986), pp. 1–22

68 copulate 500–1,000 times: Birkhead, *Promiscuity*, p. 81

68 somewhere in the middle of these two: A. H. Harcourt, P. H. Harvey, S. G. Larson and R. V. Short, 'Testis Weight, Body Weight and Breeding System in Primates' in *Nature*, 293 (1981), pp. 55–7

69 'History has not been kind to this pronouncement': Zuleyma Tang-Martínez, 'Repetition of Bateman Challenges the Paradigm' in *PNAS* (2012), pp. 11476–7

69 'eager for any female', 'fertility is seldom likely to be limited by sperm production': Bateman, 'Intra-sexual Selection in *Drosophila*', p. 364

69 combined energetics of a single ejaculate: Donald Dewsbury, 'Ejaculate Cost and Male Choice' in *The American Naturalist*, 119 (1982), pp. 601–10

69n an increased lifetime fecundity: Amy M. Worthington, Russell A. Jurenka and Clint D. Kelly, 'Mating for Male-derived Prostaglandin: A Functional Explanation for the Increased Fecundity of Mated Female Crickets?' in *Journal of Experimental Biology* (Sept. 2015)

70 recovery can take as long as 156 days: Tang-Martínez, 'Rethinking Bateman's Principles'

70 how much sperm he is prepared to spend: Nina Wedell, Matthew J. G. Gage and Geoffrey Parker, 'Sperm Competition, Male Prudence and Sperm-limited Females' in *Trends in Ecology and Evolution* (2002), pp. 313–20

70 30 per cent of the time: Cordelia Fine, *Testosterone Rex* (W. W. Norton and Co., 2017), p. 41

70 chase away soliciting females: Tang-Martínez, 'Rethinking Bateman's Principles'

71 'The labels . . . did not capture the variation': Patricia Adair Gowaty, Rebecca Steinichen and Wyatt W. Anderson, 'Indiscriminate Females and Choosy Males: Within- and Between- Species Variation in *Drosophila*' in *Evolution*, 57: 9 (2003), pp. 2037–45

71 could have got very different results: Birkhead, *Promiscuity*, pp. 197–8

71 wait longer before mating again: Tang-Martínez, 'Rethinking Bateman's Principles'

71 'important to know that Bateman's data are robust': Patricia Adair Gowaty and Brian F. Snyder, 'A Reappraisal of Bateman's Classic Study of Intrasexual Selection' in *Evolution* (The Society for the Study of Evolution), 61: 11 (2007), pp. 2457–68

72 a bad case of confirmation bias: Patricia Adair Gowaty, 'Biological Essentialism, Gender, True Belief, Confirmation Biases, and Skepticism' in *Handbook of the Psychology of Women: Vol. 1. History, Theory, and Battlegrounds* (2018), ed. by C. B. Travis and J. W. White, pp. 145–64

72 'Bateman's results are unreliable': Gowaty and Snyder, 'A Reappraisal of Bateman's Classic Study'

72 When Gowaty repeated his experiments: Patricia Adair Gowaty, Yong-Kyu Kim and Wyatt W. Anderson, 'No Evidence of Sexual Selection in a Repetition of Bateman's Classic Study of *Drosophila melanogaster*' in *PNAS*, 109 (2012), pp. 11740–5 and Thierry Hoquet, William C. Bridges, Patricia Adair Gowaty, 'Bateman's Data: Inconsistent with "Bateman's Principles"', *Ecology and Evolution*, 10: 19 (2020)

73 'Most females were uninterested in copulating more than once or twice': Robert Trivers, 'Parental Investment and Sexual Selection' in *Sexual Selection and the Descent of Man*, ed. by Bernard Campbell (Aldine-Atherton, 1972), p. 54

73 'unashamedly that it was pure bias': Tim Birkhead, 'How Stupid Not to Have Thought of That: Post-copulatory Sexual Selection' in *Journal of Zoology*, 281 (2010), pp. 78–93

74 Gowaty and her collaborator, Malin Ah-King: Malin Ah-King and Patricia Adair Gowaty, 'A Conceptual Review of Mate Choice: Stochastic Demography, Within-sex Phenotypic Plasticity, and Individual Flexibility' in *Ecology and Evolution*, 6: 14 (2016), pp. 4607–42

74 increase their reproductive fitness from promiscuous behaviour: Tang-Martínez, 'Rethinking Bateman's Principles'

74 and taught as such: ibid.

74 'evolved sex roles ultimately rest on anisogamy': Lukas Schärer, Locke Rowe and Göran Arnqvist, 'Anisogamy, Chance and the Evolution of Sex Roles' in *Trends in Ecology & Evolution*, 5 (2012), pp. 260–4

75 'God-Jesus paper': Angela Saini, *Inferior* (Fourth Estate, 2017)

75 'very political perspectives': interview with a professor of evolutionary biology at Oxford University, conducted by Jenny Easley for the book, June 2020

75 all have the power to shape their nature: Patricia Adair Gowaty, 'Adaptively Flexible Polyandry' in *Animal Behaviour*, 86 (2013), pp. 877–84

75 availability of food can cause sex roles to switch: Tang-Martínez, 'Rethinking Bateman's Principles'

Chapter Four: Fifty ways to eat your lover

77 around 125 times his mass: Matjaž Kuntner, Shichang Zhang, Matjaž Gregorič and Daiqin Li, '*Nephila* Female Gigantism Attained Through Post-maturity Molting' in *Journal of Arachnology*, 40 (2012), pp. 345–7

78 'sometimes to an extraordinary degree', 'advances', 'carries her coyness to a dangerous pitch': Charles Darwin, *The Descent of Man* (John Murray, 2nd edn, 1879; republished by Penguin Classics, 2004), pp. 314–15

78 'in the midst of his preparatory caresses': ibid.

80 on flubs were observed to be 'usual' and occur 'often': Bernhard A. Huber, 'Spider Reproductive Behaviour: A Review of Gerhardt's Work from 1911–1933, With Implications for Sexual Selection' in *Bulletin of the British Arachnological Society*, 11: 3 (1998), pp. 81–91

82 incompatible sexual agendas: Göran Arnqvist and Locke Rowe, *Sexual Conflict* (Princeton University Press, 2005)

83 before making their move: Lutz Fromhage and Jutta M. Schneider, 'Safer Sex with Feeding Females: Sexual Conflict in a Cannibalistic Spider' in *Behavioral Ecology*, 16: 2 (2004), pp. 377–82

83 can actually smell if their fancy is hungry: Luciana Baruffaldi, Maydianne C. B. Andrade, 'Contact Pheromones Mediate Male Preference in Black Widow Spiders: Avoidance of Hungry Sexual Cannibals?' in *Animal Behaviour*, 102 (2015), pp. 25–32

83 the female releases herself from her silken fetters: Alissa G. Anderson and Eileen A. Hebets, 'Benefits of Size Dimorphism and Copulatory Silk Wrapping in the Sexually Cannibalistic Nursery Web Spider, *Pisaurina mira*' in *Biology Letters*, 12 (2016)

83 giving the genital drooler an additional paternity advantage: Matjaž Gregorič, Klavdija Šuen, Ren-Chung Cheng, Simona

Kralj-Fišer and Matjaž Kuntner, 'Spider Behaviors Include Oral Sexual Encounters' in *Scientific Reports*, 6 (Nature, 2016)

84 the spider ménage à trois: Matthew H. Persons, 'Field Observations of Simultaneous Double Mating in the Wolf Spider *Rabidosa punctulata* (Araneae: Lycosidae)' in *Journal of Arachnology*, 45: 2 (2017), pp. 231–4

84 his mutilated genital plugs the female's epigynum: Daiqin Li, Joelyn Oh, Simona Kralj-Fišer and Matjaž Kuntner, 'Remote Copulation: Male Adaptation to Female Cannibalism' in *Biology Letters* (2012), pp. 512–15

84 97 per cent male survival: Gabriele Uhl, Stefanie M. Zimmer, Dirk Renner and Jutta M. Schneider, 'Exploiting a Moment of Weakness: Male Spiders Escape Sexual Cannibalism by Copulating with Moulting Females' in *Scientific Reports* (Nature, 2015)

84 The US 'evolutionist laureate': John Alcock, 'Science and Nature: Misbehavior', *Boston Review*, 1 April 2000, http://bostonreview.net/books-ideas/john-alcock-misbehavior

85 'If it occurred always, or even often': Stephen Jay Gould, 'Only His Wings Remained' in *The Flamingo's Smile: Reflections in Natural History* (W. W. Norton & Company, 1985), p. 51

85 'indiscriminate rapacity': ibid., p. 53

86 complete with mini water-weed islands: 'Life History', Fen Raft Spider Conservation [accessed 28 Jan. 2021], https://dolomedes.org.uk/index.php/biology/life_history

86 male spider's seduction routine: Shichang Zhang, Matjaž Kuntner and Daiqin Li, 'Mate Binding: Male Adaptation to Sexual Conflict in the Golden Orb-web Spider (Nephilidae: *Nephila pilipes*)' in *Animal Behaviour* 82: 6 (2011), pp. 1299–304

88 Videos of his kaleidoscopic display: Jurgen Otto, 'Peacock Spider 7 (*Maratus speciosus*)', YouTube, 2013, https://www.youtube.com/watch?v=d_yYC5r8xMI

88 judged their suitors on looks alone: Robert R. Jackson and Simon D. Pollard, 'Jumping Spider Mating Strategies: Sex Among the Cannibals in and out of Webs' in *The Evolution of*

Mating Systems in Insects and Arachnids, ed. by Jae C. Choe and Bernard J. Crespi (Cambridge University Press, 1997), pp. 340–51

91 names like 'rumble-rumps' and 'grind-revs': Madeline B. Girard, Damian O. Elias and Michael M. Kasumovic, 'Female Preference for Multi-modal Courtship: Multiple Signals are Important for Male Mating Success in Peacock Spiders' in *Proceedings of the Royal Society B*, 282 (2015); and Damian O. Elias, Andrew C. Mason, Wayne P. Maddison and Ronald R. Hoy, 'Seismic Signals in a Courting Male Jumping Spider (Araneae: Salticidae)' in *Journal of Experimental Biology* (2003), pp. 4029–39

91 as complex as any made by humans: Damian O. Elias, Wayne P. Maddison, Christina Peckmezian, Madeline B. Girard, Andrew C. Mason, 'Orchestrating the Score: Complex Multimodal Courtship in the *Habronattus coecatus* Group of *Habronattus* Jumping Spiders (Araneae: Salticidae)' in *Biological Journal of the Linnean Society*, 105: 3 (2012), pp. 522–47

92 starts his seduction routine with jazz hands: Jackson and Pollard, 'Jumping Spider Mating Strategies: Sex Among Cannibals in and out of Webs'; and David L. Clark and George W. Uetz, 'Morph-independent Mate Selection in a Dimorphic Jumping Spider: Demonstration of Movement Bias in Female Choice Using Video-controlled Courtship Behaviour' in *Animal Behaviour*, 43: 2 (1992), pp. 247–54

93 these quivers can be co-opted: Marie E. Herberstein, Anne E. Wignall, Eileen A. Hebets and Jutta M. Schneider, 'Dangerous Mating Systems: Signal Complexity, Signal Content and Neural Capacity in Spiders' in *Neuroscience & Biobehavioral Reviews*, 46: 4 (2014), pp. 509–18

94 they suck her dry: M. Salomon, E. D. Aflalo, M. Coll and Y. Lubin, 'Dramatic Histological Changes Preceding Suicidal Maternal Care in the Subsocial Spider *Stegodyphus lineatus* (Araneae: Eresidae)' in *Journal of Arachnology*, 43: 1 (2015), pp. 77–85

94 'ferocity of the female': Darwin, *The Descent of Man*, p. 315

94 excessive burdens of spider motherhood: Gustavo Hormiga, Nikolaj Scharff and Jonathan A. Coddington, 'The Phylogenetic Basis of Sexual Size Dimorphism in Orb-weaving Spiders (Araneae, Orbiculariae)' in *Systematic Biology*, 49: 3 (2000), pp. 435–62

95 'Spider Bites Australian Man on Penis Again': 'Spider Bites Australian Man on Penis Again', BBC News, 28 Sept. 2016, https://www.bbc.co.uk/news/world-australia-37481251

96 'Gerhardt's position number 3': L. M. Foster, 'The Stereotyped Behaviour of Sexual Cannibalism in *Latrodectus-Hasselti Thorell* (Araneae, Theridiidae), the Australian Redback Spider' in *Australian Journal of Zoology*, 40 (1992), pp. 1–11

96 'with legs flailing': ibid.

96 'small blobs of white substances': ibid.

97 their one shot at sex hits the target: Maydianne C. B. Andrade, 'Sexual Selection for Male Sacrifice in the Australian Redback Spider' in *Science*, 271 (1996), pp. 70–72

97 concrete supporting evidence has been elusive: Jutta M. Schneider, Lutz Fromhage and Gabriele Uhl, 'Fitness Consequences of Sexual Cannibalism in Female *Argiope bruennichi*' in *Behavioral Ecology and Sociobiology*, 55 (2003), pp. 60–64

97 something uniquely nutritious about eating the male: Steven K. Schwartz, William E. Wagner, Jr. and Eileen A. Hebets, 'Males Can Benefit from Sexual Cannibalism Facilitated by Self-sacrifice' in *Current Biology*, 26 (2016), pp. 1–6

98 'a negative stereotype of sexually aggressive females': Liam R. Dougherty, Emily R. Burdfield-Steel and David M. Shuker, 'Sexual Stereotypes: the Case of Sexual Cannibalism' in *Animal Behaviour* (2013), pp. 313–22

Chapter Five: Love is a battlefield

99 'after a period of time . . . the tiny fetuses': Carl G. Hartman, *Possums* (University of Texas at Austin, 1952), p. 84

100 a temporary third vagina: William John Krause, *The Opossum: Its Amazing Story* (Department of Pathology and Anatomical Sciences, School of Medicine, University of Missouri, 2005)

100 widespread use of penis morphology in taxonomy: William G. Eberhard, 'Postcopulatory Sexual Selection: Darwin's Omission and its Consequences', *PNAS*, 6 (2009), pp. 10025–32

101 humdrum fleshy tube by comparison: Menno Schilthuizen, *Nature's Nether Regions* (Viking, 2014), p. 5

101 under some powerful selection forces: Eberhard, 'Postcopulatory Sexual Selection'

102 spring-clean the female's reproductive tract: J. K. Waage, 'Dual Function of the Damselfly Penis: Sperm Removal and Transfer' in *Science*, 203 (1979), pp. 916–18

102n 'deeper and more vigorous penile thrusting': Gordon G. Gallup Jr., Rebecca L. Burch, Mary L. Zappieri, Rizwan A. Parvez, Malinda L. Stockwell and Jennifer A. Davis, 'The Human Penis as a Semen Displacement Device' in *Evolution and Human Behaviour*, 24: 4 (July 2003), pp. 277–89

103 engaged in heated debate: William G. Eberhard, 'Rapid Divergent Evolution of Genitalia' in *The Evolution of Primary Sexual Characters in Animals*, ed. by Alex Córdoba-Aguilar and Janet L. Leonard (Oxford University Press, 2010), pp. 40–78; and Paula Stockley and David J. Hosken, 'Sexual Selection and Genital Evolution' in *Trends in Ecology & Evolution*, 19: 2 (2014), pp. 87–93

103 'one of evolutionary biology's greatest enigmas': Malin Ah-King, Andrew B. Barron and Marie E. Herberstein, 'Genital Evolution: Why Are Females Still Understudied?' in *PLoS Biology*, 12: 5 (2014), pp. 1–7

103 'scientifically unstoppable': Richard O. Prum, *The Evolution of Beauty* (Anchor Books, 2017), p. 162

106 corkscrews counter-clockwise: Kevin G. McCracken, Robert E. Wilson, Pamela J. McCracken and Kevin P. Johnson, 'Are Ducks Impressed by Drakes' Display?' in *Nature*, 413: 128 (2001)

106 penis explodes out of his cloaca at 75 mph: Patricia L. R. Brennan, Christopher J. Clark and Richard O. Prum, 'Explosive Eversion and Functional Morphology of Waterfowl Penis Supports Sexual Conflict in Genitalia' in *Proceedings of the Royal Society B* (2010), pp. 1309–14

106 a result of sperm competition: McCracken, Wilson, McCracken and Johnson, 'Are Ducks Impressed by Drakes' Display?'

107n have suggested that human rape is biologically determined by Darwinism: Craig Palmer and Randy Thornhill, *A Natural History of Rape: Biological Bases of Coercion* (MIT Press, 2000)

107n forced a strict avoidance of the human term: Patricia Adair Gowaty, 'Forced or Aggressively Coerced Copulation' in *Encyclopedia of Animal Behaviour* (Elsevier, 2010), p. 760

108 the result of an escalating arms race: Brennan, Clark and Prum, 'Explosive Eversion and Functional Morphology'

109 the female's vagina correspondingly simple: Patricia L. R. Brennan, Richard O. Prum, Kevin G. McCracken, Michael D. Sorenson, Robert E. Wilson and Tim R. Birkhead, 'Coevolution of Male and Female Genital Morphology in Waterfowl' in *PLoS One*, 2: 5 (2007)

111 the penis eventually disappeared: Patricia L. R. Brennan, 'Genital Evolution: Cock-a-Doodle-Don't' in *Current Biology*, 23: 12 (2013), pp. 523–5

111 impossible to fertilize the female without her consent: Gowaty, 'Forced or Aggressively Coerced Copulation', pp. 759–63

111 compete with one another to provide the best care: Prum, *The Evolution of Beauty*, pp. 179–81

112 'Too often, the female is assumed to be an invariant container': Ah-King, Barron and Herberstein, 'Genital Evolution: Why Are Females Still Understudied?'

112 genital loss is apparently quite common with earwigs: Yoshitaka Kamimura and Yoh Matsuo, 'A "Spare" Compensates for the Risk of Destruction of the Elongated Penis of Earwigs (Insecta: Dermaptera)' in *Naturwissenschaften* (2001), pp. 468–71

112 used his lengthy virga like a chimney sweep's brush: 'Last-male Paternity of *Euborellia plebeja*, an Earwig with Elongated

Genitalia and Sperm-removal Behaviour' in *Journal of Ethology* (2005), pp. 35–41

112 'Thus, females seem to beat males': Yoshitaka Kamimura, 'Promiscuity and Elongated Sperm Storage Organs Work Cooperatively as a Cryptic Female Choice Mechanism in an Earwig' in *Animal Behaviour*, 85 (2013), pp. 377–83

113 'influenced by male-centred outlooks': William G. Eberhard, 'Inadvertent Machismo?' in *Trends in Ecology & Evolution*, 5: 8 (1990) p. 263

113 deemed to have no influence: Marlene Zuk, *Sexual Selections: What We Can and Can't Learn about Sex from Animals* (University of California Press, 2002), p. 82

113 'the rules of the game': William G. Eberhard, *Female Control: Sexual Selection by Cryptic Female Choice* (Princeton University Press, 1996)

113 yet even this advocate: Patricia L. R. Brennan, 'Studying Genital Coevolution to Understand Intromittent Organ Morphology' in *Integrative and Comparative Biology*, 56: 4 (2016), pp. 669–81

114 'aquatic bonobos': Takeshi Furuichi, Richard Connor and Chie Hashimoto, 'Non-conceptive Sexual Interactions in Monkeys, Apes and Dolphins' in *Primates and Cetaceans: Field Research and Conservation of Complex Mammalian Societies*, ed. by Leszek Karczmarski and Juichi Yamagiwa (Springer, 2014), p. 390

114n dolphin named Zafar began sexually harassing: 'Sexually Frustrated Dolphin Named Zafar Sexually Terrorizes Tourists on a French Beach' (*Telegraph*, 27 August 2018), https://www.telegraph.co.uk/news/2018/08/27/swimming-banned-french-beach-sexually-frustrated-dolphin-named/

115 unwanted suitor gets sent up a blind alley: Dara N. Orbach, Diane A. Kelly, Mauricio Solano and Patricia L. R. Brennan, 'Genital Interactions During Simulated Copulation Among Marine Mammals' in *Proceedings of the Royal Society B*, 284: 1864 (2017)

115 females grow bigger brains in order to outwit their aggressors: Séverine D. Buechel, Isobel Booksmythe, Alexander Kotrschal, Michael D. Jennions and Niclas Kolm, 'Artificial Selection on Male Genitalia Length Alters Female Brain Size' in *Proceedings of the Royal Society B*, 283: 1843 (2016)

116 using a deli counter meat slicer: Patricia L. R. Brennan and Dara N. Orbach, 'Functional Morphology of the Dolphin Clitoris' in *The FASEB journal*, 3: S1 (2019), p. 10.4

117 'this new and useless part', 'healthy': Helen E. O'Connell, Kalavampara V. Sanjeevan and John M. Hutson, 'Anatomy of the Clitoris' in *Journal of Urology*, 174: 4 (2005), p. 1189

117 only to be found in hermaphrodites: Schilthuizen, *Nature's Nether Regions*, p. 74

117 remove the clitoral label: Adele E. Clarke and Lisa Jean Moore, 'Clitoral Conventions and Transgressions: Graphic Representations in Anatomy Texts' in *Feminist Studies*, 21: 2 (1995), p. 271

117 'small version of the penis': O'Connell, Sanjeevan and Hutson, 'Anatomy of the Clitoris'

117n unlikely investigative authority on female reproductive anatomy: M. M. Mortazavi, N. Adeeb, B. Latif, K. Watanabe, A. Deep, C. J. Griessenauer, R. S. Tubbs and T. Fukushima, 'Gabriele Falloppio (1523–1562) and His Contributions to the Development of Medicine and Anatomy' in *Child's Nervous System* (2013) pp. 877–80

117n 'trumpets of the uterus': Çağatay Öncel, 'One of the Great Pioneers of Anatomy: Gabriele Falloppio (1523–1562)' in *Bezmialem Science*, 123 (2016)

117n 'appeal to women': 'Gabriele Falloppio', Whonamedit? A Dictionary of Medical Eponyms, http://www.whonamedit.com/doctor.cfm/2288.html

118 first detailed anatomy of the human clitoris: Helen O'Connell, 'Anatomical Relationship Between Urethra and Clitoris' in *Journal of Urology*, 159: 6 (1998), pp. 1892–7

119 had his titillators experimentally shortened: Nadia S. Sloan and Leigh W. Simmons, 'The Evolution of Female Genitalia' in *Journal of Evolutionary Biology* (2019), pp. 1–18

119 decides if a male gets to fertilize her eggs: Eberhard, *Female Control*

119 finishes with the farmer sitting on her back: Víctor Poza Moreno, 'Stimulation During Insemination: The Danish Perspective', Pig333.com Professional Pig Community, 15 Sept. 2011, https://www.pig333.com/articles/stimulation-during-insemination-the-danish-perspective_4812/

119 They need help: Teri J. Orr and Virginia Hayssen, *Reproduction in Mammals: The Female Perspective* (Johns Hopkins University Press, 2017)

119 significantly accelerates their passage to the egg: David A. Puts, Khytam Dawood and Lisa L. M. Welling, 'Why Women Have Orgasms: An Evolutionary Analysis' in *Archives of Sexual Behaviour*, 41: 5 (2012), pp. 1127–43

119 A study of captive Japanese macaques: Monica Carosi and Alfonso Troisi, 'Female Orgasm Rate Increases With Male Dominance in Japanese Macaques' in *Animal Behaviour* (1998), pp. 1261–6

120 cryptic means of selecting a high-quality sire: Puts, Dawood and Welling, 'Why Women Have Orgasms'

120 'although recognized for more than fifty years': Orr and Hayssen, *Reproduction in Mammals*, p. 115

121 awaken her from her lifeless slumber: Emily Martin, 'The Egg and the Sperm: How Science Has Constructed a Romance Based on Stereotypical Male–Female Roles' in *Signs* (University of Chicago Press), 16: 3 (1991), pp. 485–501

121 preferred the sperm of a random male: John L. Fitzpatrick, Charlotte Willis, Alessandro Devigili, Amy Young, Michael Carroll, Helen R. Hunter and Daniel R. Brison, 'Chemical Signals from Eggs Facilitate Cryptic Female Choice in Humans' in *Proceedings of the Royal Society B*, 287: 1928 (2020)

Chapter Six: *Madonna no more*

122 'Woman seems to differ from man in mental disposition':
Charles Darwin, *The Descent of Man, and Selection in Relation
to Sex* (John Murray, 2nd edn, 1879; republished by Penguin
Classics, 2004), p. 629

125 driven him away, quite unsentimentally: Adam Davis, '*Aotus
nigriceps* Black-headed Night Monkey', Animal Diversity Web
(University of Michigan), https://animaldiversity.org/accounts/
Aotus_nigriceps/

125 only one in ten mammalian species: David J. Hosken and
Thomas H. Kunz, 'Male Lactation: Why, Why Not and Is It
Care?' in *Trends in Ecology & Evolution*, 24: 2 (2008), pp.
80–5

125n 'it is odd that no case of male lactation has evolved': John
Maynard Smith, *The Evolution of Sex* (Cambridge University
Press, 1978)

125n could have stimulated the mammary tissue: C. M. Francis,
Edythe L. P. Anthony, Jennifer A. Brunton, Thomas H. Kunz,
'Lactation in Male Fruit Bats' in *Nature* (1994), pp. 691–2

125n It is likely the same story with the inbred domesticated sheep:
Hosken and Kunz, 'Male Lactation'

126n remarkably womb-like: Camilla M. Whittington, Oliver
W. Griffith, Weihong Qi, Michael B. Thompson and Anthony
B. Wilson, 'Seahorse Brood Pouch Transcriptome Reveals Com-
mon Genes Associated with Vertebrate Pregnancy' in *Molecular
Biology and Evolution*, 32: 12 (2015), pp. 3114–31

128 both retain the brain architecture: Eva K. Fischer, Alexandre
B. Roland, Nora A. Moskowitz, Elicio E. Tapia, Kyle Summers,
Luis A. Coloma and Lauren A. O'Connell, 'The Neural Basis of
Tadpole Transport in Poison Frogs' in *Proceedings of the Royal
Society B*, 286 (2019)

128 can be transformed into doting dads: Z. Wu, A. E. Autry, J. F.
Bergan, M. Watabe-Uchida and Catherine G. Dulac, 'Galanin

Neurons in the Medial Preoptic Area Govern Parental Behaviour' in *Nature*, 509 (2014), pp. 325–30

131 'Mothers were viewed as one-dimensional automatons': Sarah Blaffer Hrdy, *Mother Nature* (Ballantine Books, 1999), p. 27

132 'Most adult females in most animal populations': Margo Wilson and Martin Daly, *Sex, Evolution and Behaviour* (Thompson/Duxbury Press, 1978)

133 'Observational Study of Behaviour: Sampling Methods': Jeanne Altmann, 'Observational Study of Behaviour: Sampling Methods' in *Behaviour*, 4 (1974), pp. 227–67

133 'inadvertently, one of the greatest feminist papers of all time': Interview with Dr Rebecca Lewis, anthropology professor, University of Texas at Austin, March 2016

133 ' "the home economics" of animal behaviour': Hrdy, *Mother Nature*, p. 46

134 'dual career' mother spending 70 per cent of each day 'making a living': Jeanne Altmann, *Baboon Mothers and Infants* (Harvard University Press, 1980), p. 6

134 'Vee's first infant, Vicki': ibid., pp. 208–9

134 up to 60 per cent higher: Hrdy, *Mother Nature*, p. 155

135 'psychological weapons': Robert L. Trivers, 'Parent–Offspring Conflict' in *American Zoology*, 14 (1974), pp. 249–64

135 only two are likely to survive: Hrdy, *Mother Nature*, p. 334

137 self-sufficient and socially integrated juveniles: Joan B. Silk, Susan C. Alberts and Jeanne Altmann, 'Social Bonds of Female Baboons Enhance Infant Survival' in *Science*, 302 (2003), pp. 1231–4

137 make mothers more vulnerable to disease: Dario Maestripieri, 'What Cortisol Can Tell Us About the Costs of Sociality and Reproduction Among Free-ranging Rhesus Macaque Females on Cayo Santiago' in *American Journal of Primatology*, 78 (2016), pp. 92–105

137 higher levels of abusive behaviour during the postpartum period: Linda Brent, Tina Koban and Stephanie Ramirez, 'Abnormal, Abusive, and Stress-related Behaviours in Baboon

Mothers' in *Society of Biological Psychiatry*, 52: 11 (2002), pp. 1047–56

137 crushing their infants on the ground: Dario Maestripieri, 'Parenting Styles of Abusive Mothers in Group-living Rhesus Macaques' in *Animal Behaviour*, 55: 1 (1998), pp. 1–11

137 more likely to mistreat their own young: Maestripieri, 'Early Experience Affects the Intergenerational Transmission of Infant Abuse in Rhesus Monkeys' in *PNAS*, 102: 27 (2005), pp. 9726–9

138 they can gain much-needed assistance: Silk, Alberts and Altmann, 'Social Bonds of Female Baboons Enhance Infant Survival'

138 ability to forge strong and enduring social bonds: Joan B. Silk, Jacinta C. Beehner, Thore J. Bergman, Catherine Crockford, Anne L. Engh, Liza R. Moscovice, Roman M. Wittig, Robert M. Seyfarth and Dorothy L. Cheney, 'The Benefits of Social Capital: Close Social Bonds Among Female Baboons Enhance Offspring Survival' in *Proceedings of the Royal Society B*, 276 (2009), pp. 3099–104

138 low-ranking females had more sons than daughters: Jeanne Altmann, Glenn Hausfater and Stuart A. Altmann, 'Determinants of Reproductive Success in Savannah Baboons, *Papio cynocephalus*' in *Reproductive Success: Studies of Individual Variation in Contrasting Breeding Systems*, ed. by Tim H. Clutton-Brock (University of Chicago Press, 1988), pp. 403–18

139n José Tella discovered this in 2001: J. L. Tella, 'Sex Ratio Theory in Conservation Biology' in *Ecology and Evolution* (2001), pp. 76–7

140 Rich milk may help sons bulk up faster: Katherine Hinde, 'Richer Milk for Sons But More Milk for Daughters: Sex-biased Investment during Lactation Varies with Maternal Life History in Rhesus Macaques' in *American Journal of Human Biology*, 21: 4 (2009), pp. 512–19

140 their method is strategic abortion: Hrdy, *Mother Nature*, p. 330

140 this particular trigger for abortion: Eila K. Roberts, Amy Lu, Thore J. Bergman and Jacinta C. Beehner, 'A Bruce Effect in Wild Geladas' in *Science*, 335: 6073 (2012), pp. 1222–5

141 trigger her embryo-in-waiting to emerge: Hrdy, *Mother Nature*, p. 129

142 bonobo mothers act as matchmakers for their sons: Martin Surbeck, Christophe Boesch, Catherine Crockford, Melissa Emery Thompson, Takeshi Furuichi, Barbara Fruth, Gottfried Hohmann, Shintaro Ishizuka, Zarin Machanda, Martin N. Muller, Anne Pusey, Tetsuya Sakamaki, Nahoko Tokuyama, Kara Walker, Richard Wrangham, Emily Wroblewski, Klaus Zuberbühler, Linda Vigilant and Kevin Langergraber, 'Males with a Mother Living in their Group Have Higher Paternity Success in Bonobos But Not Chimpanzees' in *Current Biology*, 29: 10 (2019), pp. 341–57

142 'plodding constants': Hrdy, *Mother Nature*, p. 83

144 offers a unique opportunity to study maternal behaviour: S. Smout, R. King and P. Pomeroy, 'Environment-sensitive Mass Changes Influence Breeding Frequency in a Capital Breeding Marine Top Predator' in *Journal of Animal Ecology*, 88: 2 (2019), pp. 384–96

144 recent medical paper documenting a unique case of 'seal buttock': Timur Kouliev and Victoria Cui, 'Treatment and Prevention of Infection Following Bites of the Antarctic Fur Seal (*Arctocephalus gazella*)' in Open Access Emergency Medicine (2015), pp. 17–20

145 When chimpanzees groom one another: C. Crockford, R. M. Wittig, K. Langergraber, T. E. Ziegler, K. Zuberbühler and T. Deschner, 'Urinary Oxytocin and Social Bonding in Related and Unrelated Wild Chimpanzees' in *Proceedings of the Royal Society B*, 280: 1755 (2013)

145 when I gaze at my pet dog: Miho Nagasawa, Shohei Mitsui, Shiori En, Nobuyo Ohtani, Mitsuaki Ohta, Yasuo Sakuma, Tatsushi Onaka, Kazutaka Mogi and Takefumi Kikusui,

'Oxytocin-gaze Positive Loop and the Coevolution of Human–Dog Bonds' in *Science*, 348 (2015), pp. 333–6

145 linking those areas to the dopamine reward system: Lane Strathearn, Peter Fonagy, Janet Amico and P. Read Montague, 'Adult Attachment Predicts Maternal Brain and Oxytocin Response to Infant Cues' in *Neuropsychopharmacology*, 34 (2009), pp. 2655–66

146 priming her to valiantly defend her offspring: Jennifer Hahn-Holbrook, Julianne Holt-Lunstad, Colin Holbrook, Sarah M. Coyne and E. Thomas Lawson, 'Maternal Defense: Breast Feeding Increases Aggression by Reducing Stress' in *Psychological Science*, 22: 10 (2011), pp. 1288–95

146 she loses up to 40 per cent of her body weight: M. A. Fedak and S. S. Anderson, 'The Energetics of Lactation: Accurate Measurements from a Large Wild Mammal, the Grey Seal (*Halichoerus grypus*)' in *Journal of Zoology*, 198: 2 (1982), pp. 473–9

147 it was comparable to a non-breeding female: Kelly J. Robinson, Sean D. Twiss, Neil Hazon and Patrick P. Pomeroy, 'Maternal Oxytocin Is Linked to Close Mother–Infant Proximity in Grey Seals (*Halichoerus grypus*)' in *PLoS One*, 10: 12 (2015), pp. 1–17

147 If the mother gets distracted during this critical period: ibid.

148 resulted in the fattest pups: Kelly J. Robinson, Neil Hazon, Sean D. Twiss, Patrick P. Pomeroy, 'High Oxytocin Infants Gain More Mass with No Additional Maternal Energetic Costs in Wild Grey Seals (*Halichoerus grypus*)' in *Psychoneuroendocrinology*, 110 (2019)

148 those with attentive mothers: James K. Rilling and Larry J. Young, 'The Biology of Mammalian Parenting and its Effect on Offspring Social Development' in *Science*, 345: 6198 (2014), pp. 771–6

148 failed to forge lifelong sexual bonds: Allison M. Perkeybile, C. Sue Carter, Kelly L. Wroblewski, Meghan H. Puglia, William M. Kenkel, Travis S. Lillard, Themistoclis Karaoli, Simon

G. Gregory, Niaz Mohammadi, Larissa Epstein, Karen L. Bales and Jessica J. Connell, 'Early Nurture Epigenetically Tunes the Oxytocin Receptor' in *Psychoneuroendocrinology*, 99 (2019), pp. 128–36

149 unique ability to recognize different sensory cues: Lane Strathearn, Jian Li, Peter Fonagy and P. Read Montague, 'What's in a Smile? Maternal Brain Responses to Infant Facial Cues' in *Pediatrics*, 122: 1 (2008), pp. 40–51

149 activation in the area associated with unfairness, pain and disgust: Strathearn, Fonagy, Amico and Montague, 'Adult Attachment Predicts Maternal Brain and Oxytocin Response'

150 eventually she's as attentive as a birth mother: Hrdy, *Mother Nature*, p. 151

150 Adoption has been recorded in at least 120 mammals: Teri J. Orr and Virginia Hayssen, *Reproduction in Mammals: The Female Perspective* (Johns Hopkins University Press, 2017)

153 friends – both male and female – are equally, if not more, important: Andrea L. Baden, Timothy H. Webster and Brenda J. Bradley, 'Genetic Relatedness Cannot Explain Social Preferences in Black-and-white Ruffed Lemurs, *Varecia variegata*' in *Animal Behaviour*, 164 (2020), pp. 73–82

154 around 10–13 million calories to rear a human: Hrdy, *Mother Nature*, p. 177

154 'greater intellectual vigour and power of invention of man': Charles Darwin, *The Descent of Man, and Selection in Relation to Sex* (reprinted Gale Research, 1974; first published 1874), p. 778

154 'their mothers are breeding at a much faster pace than great apes': 'The Evolution of Motherhood', *Nova*, 26 Oct. 2009, https://www.pbs.org/wgbh/nova/article/evolution-motherhood/

155 'greater tenderness and less selfishness': Darwin, *The Descent of Man* (John Murray, 2nd edn, 1879; republished by Penguin Classics, 2004), p. 629

Chapter Seven: Bitch eat bitch

156 'the law of battle': Charles Darwin, *The Descent of Man, and Selection in Relation to Sex* (John Murray, 2nd edn, 1879; republished by Penguin Classics, 2004), pp. 561–75

156 'It is certain that amongst almost all animals': ibid., p. 246

157 'desperate conflicts during the season of love': ibid., p. 561

157 'waste of vital power', 'of any special use, but simply in inheritance': ibid., p. 566

157 'When biologists talk about the "battle of the sexes"': Roxanne Khamsi, 'Male Antelopes Play Hard to Get' in *New Scientist*, 29 Nov. 2007, https://www.newscientist.com/article/dn12979-male-antelopes-play-hard-to-get-/

158 she often winds up being mounted by the fraudster: Wiline M. Pangle and Jakob Bro-Jørgensen, 'Male Topi Antelopes Alarm Snort Deceptively to Retain Females for Mating' in *The American Naturalist* (2010), pp. 33–9

158 males collapsing with exhaustion: Khamsi, 'Male Antelopes Play Hard to Get'

158 sperm supplies are far from cheap and limitless: Richard Dawkins, *The Selfish Gene* (Oxford University Press, 2nd edn, 1989; 1st edn, 1976)

158 charge top studs in the act of mounting other females: Jakob Bro-Jørgensen, 'Reversed Sexual Conflict in a Promiscuous Antelope' in *Current Biology*, 17 (2007), pp. 2157–61

159 deliberately choose the females they have mated with the least: ibid.

159 opposite sexual conflicts may occur more commonly than we think: 'Male Topi Antelope's Sex Burden', BBC News, 28 Nov. 2007, http://news.bbc.co.uk/1/mobile/sci/tech/7117498.stm

159 'spiteful strategy': Diane M. Doran-Sheehy, David Fernandez and Carola Borries, 'The Strategic Use of Sex in Wild Female Western Gorillas' in *American Journal of Primatology*, 71 (2009), pp. 1011–20

159 monopolize the silverback's sperm and resources: Tara S. Stoinski, Bonne M. Perdue and Angela M. Legg, 'Sexual Behavior in Female Western Lowland Gorillas (*Gorilla gorilla gorilla*): Evidence for Sexual Competition' in *American Journal of Primatology*, 71 (2009), pp. 587–93

159 'few anomalous cases', 'reversed', 'properly belong to the males': Darwin, *The Descent of Man* (1871)

160 combative potential of females was largely ignored by science: Paula Stockley and Jakob Bro-Jørgensen, 'Female Competition and its Evolutionary Consequences in Mammals' in *Biological Review*, 86 (2011), pp. 341–66

160 female vocalizations were the result of a 'hormonal imbalance': K. A. Hobson and S. G. Sealy, 'Female Song in the Yellow Warbler' in *Condor*, 92 (1990), pp. 259–61; and Rachel Mundy, *Animal Musicalities: Birds, Beasts, and Evolutionary Listening* (Wesleyan University Press, 2018), p. 38

160 '"complex vocalizations by male birds during the breeding season"': Clive K. Catchpole and Peter J. B. Slater, *Bird Song: Biological Themes and Variations* (Cambridge University Press, 2005)

161 71 per cent of female songbirds sing: Karan J. Odom, Michelle L. Hall, Katharina Riebel, Kevin E. Omland and Naomi E. Langmore, 'Female Song is Widespread and Ancestral in Songbirds' in *Nature Communications*, 5 (2014), p. 3379

161 comprising 60 per cent of all known birds: Oliver L. Austen, 'Passeriform', Britannica, https://www.britannica.com/animal/passeriform

161 so are easily mistaken for noisy males: Naomi Langmore, 'Quick Guide to Female Birdsong' *Current Biology*, 30 (2020), pp. R783–801

162 'I could hardly miss that anomaly': Keiren McLeonard, 'Aussie Birds Prove Darwin Wrong', *ABC*, 5 March 2014, https://www.abc.net.au/radionational/programs/archived/bushtelegraph/female-birds-hit-the-high-notes/5298150

162 evolved in Australia some forty-seven million years ago: Carl H. Oliveros et al., 'Earth History and the Passerine Superradiation' in *PNAS*, 116: 16 (2019), pp. 7916–25

162 earliest female songbirds were, indeed, a bunch of raucous divas: Odom, Hall, Riebel, Omland and Langmore, 'Female Song is Widespread and Ancestral in Songbirds'

163 can be induced to start singing: Hobson and Sealy, 'Female Song in the Yellow Warbler'

163 The concept of social selection was developed: Mary Jane West-Eberhard, 'Sexual Selection, Social Competition, and Evolution' in *Proceedings of the American Philosophical Society* (1979), pp. 222–34

164 could all be explained by the broader category of social, if not sexual, selection: Mary Jane West-Eberhard, 'Sexual Selection, Social Competition, and Speciation' in *The Quarterly Review of Biology*, 58: 2 (1983), pp. 155–83

164 don't see the need to invite yet another form: Tim H. Clutton-Brock, 'Sexual Selection in Females' in *Animal Behaviour* (2009), pp. 3–11

164 bright plumage and ornamentation in female birds: Trond Amundsen, 'Why Are Female Birds Ornamented?' in *Trends in Ecology & Evolution*, 15: 4 (2000), pp. 149–55

164 Darwin's narrow focus has 'clouded our view': Joseph A. Tobias, Robert Montgomerie and Bruce E. Lyon, 'The Evolution of Female Ornaments and Weaponry: Social Selection, Sexual Selection and Ecological Competition' in *Philosophical Transactions of the Royal Society B*, 367 (2012), pp. 2274–93

164 fight over resources related to fecundity and parenting: ibid.

165 'Defence and aggression in the hen is accomplished with the beak': D. W. Rajecki, 'Formation of Leap Orders in Pairs of Male Domestic Chickens' in *Aggressive Behavior*, 14: 6 (1988), pp. 425–36

166 an altercation with a more powerful female academic: Jack El-Hai, 'The Chicken-hearted Origins of the "Pecking Order"' in *Discover*, 5 July 2016, https://www.discovermagazine.com/planet-earth/the-chicken-hearted-origins-of-the-pecking-order

166 'They put a lot at stake': Marlene Zuk, *Sexual Selections: What We Can and Can't Learn about Sex from Animals* (University of California Press, 2002)

166 'Females are not innately disposed to organize into hierarchies': Virginia Abernethy, 'Female Hierarchy: An Evolutionary Perspective' in *Female Hierarchies*, ed. by Lionel Tiger and Heather T. Fowler (Beresford Book Service, 1978)

167 'sinister prerogative to interfere in the reproduction of other females': Sarah Blaffer Hrdy, *The Woman That Never Evolved* (Harvard University Press, 1981), p. 109

167 'obsessed with signs of status differences or disrespect': Susan Sperling, 'Baboons with Briefcases: Feminism, Functionalism, and Sociobiology in the Evolution of Primate Gender' in *Signs*, 17: 1 (1991), p. 18

167 one of the driving forces behind the increase in brain size: Richard Gray, 'Why Meerkats and Mongooses Have a Cooperative Approach to Raising their Pups', *Horizon: The EU Research and Innovation Magazine*, 27 June 2019, https://ec.europa.eu/research-and-innovation/en/horizon-magazine/why-meerkats-and-mongooses-have-cooperative-approach-raising-their-pups

168 single dominant female monopolizing 80 per cent of the breeding: Andrew J. Young and Tim Clutton-Brock, 'Infanticide by Subordinates Influences Reproductive Sharing in Cooperatively Breeding Meerkats' in *Biology Letters*, 2 (2006), pp. 385–7

168 'cooperative breeding': Tim Clutton-Brock, *Mammal Societies* (Wiley, 2016)

168 zero-tolerance policy for breeding subordinates: Sarah J. Hodge, A. Manica, T. P. Flower and T. H. Clutton-Brock, 'Determinants of Reproductive Success in Dominant Female Meerkats' in *Journal of Animal Ecology*, 77 (2008), pp. 92–102

171 'The Kalahari is an amoral, unregulated market force': A. A. Gill, *AA Gill is Away* (Simon & Schuster, 2007), pp. 36–7

172 take up wet-nursing duties for their murderous mother's babies: K. J. MacLeod, J. F. Nielsen and T. H. Clutton-Brock, 'Factors Predicting the Frequency, Likelihood and Duration of Allonursing in the Cooperatively Breeding Meerkat' in *Animal Behaviour*, 86: 5 (2013), pp. 1059–67

172 a form of punishment or 'rent' to be paid: 'Infanticide Linked to Wet-nursing in Meerkats', *Science Daily*, 7 Oct. 2013, https://www.sciencedaily.com/releases/2013/10/131007122558.htm

173 of 248 recorded litters 106 failed to emerge: Young and Clutton-Brock, 'Infanticide by Subordinates'

173 one in five probability of being killed by another meerkat: José María Gómez, Miguel Verdú, Adela González-Megías and Marcos Méndez, 'The Phylogenetic Roots of Human Lethal Violence' in *Nature*, 538 (2016), pp. 233–7

173 only one in every six or seven meerkats: Gray, 'Why Meerkats and Mongooses Have a Cooperative Approach'

174 most reproductively successful terrestrial animal: Daniel Elsner, Karen Meusemann and Judith Korb, 'Longevity and Transposon Defense, the Case of Termite Reproductives' in *PNAS* (2018), pp. 5504–9

175 rendered sterile and kept in their lowly castes: Takuya Abe and Masahiko Higashi, 'Macrotermes', Science Direct (2001) https://www.sciencedirect.com/topics/biochemistry-genetics-and-molecular-biology/macrotermes

175 one mammal society that's been classed as eusocial: F. M. Clarke and C. G. Faulkes, 'Dominance and Queen Succession in Captive Colonies of the Eusocial Naked Mole-rat, *Heterocephalus glaber*' in *Proceedings of the Royal Society B*, 264: 1384 (1997), pp. 993–1000

175 'went missing': Interview with Chris Faulkes, 28 Sept. 2020

177 might also be the reason they don't get cancer: Xiao Tian, Jorge Azpurua, Christopher Hine, Amita Vaidya, Max Myakishev-Rempel, Julia Ablaeva, Zhiyong Mao, Eviatar Nevo, Vera Gorbunova and Andrei Seluanov, 'High-molecular-mass

Hyaluronan Mediates the Cancer Resistance of the Naked Mole-rat' in *Nature*, 499 (2013), pp. 346–9

177 of particular interest to life-hacking labs in Silicon Valley: Brady Hartman, 'Google's Calico Labs Announces Discovery of a "Non-aging Mammal"', Lifespan.io, 29 Jan. 2018, https://www.lifespan.io/news/non-aging-mammal/ [accessed Dec. 2020]; and Rochelle Buffenstein, 'The Naked Mole-rat: A New Long-living Model for Human Aging Research' in *The Journals of Gerontology: Series A*, 60: 11 (2005), pp. 1369–77

177 an eye-watering twenty-seven was recorded in one litter: Chris Faulkes, 'Animal Showoff', July 2014 (YouTube, 15 April 2015), https://www.youtube.com/watch?v=6VmxP7nDQnM

177 'gelatinous': 'Naked Mole-rat (*Heterocephalus glaber*) Fact Sheet: Reproduction & Development', San Diego Zoo Wildlife Alliance Library, https://ielc.libguides.com/sdzg/factsheets/naked-mole-rat/reproduction

177 reared more than nine hundred pups: Chris Faulkes, 'Animal Showoff'

178 eating the queen's chemically controlled faeces: Daniel E. Rozen, 'Eating Poop Makes Naked Mole-rats Motherly' in *Journal of Experimental Biology*, 221: 21 (2018)

179 subordinate will be left to squeeze under her: Clarke and Faulkes, 'Dominance and Queen Succession'

179 necessary to maintain the colony's sexual suppression: C. G. Faulkes and D. H. Abbot, 'Evidence that Primer Pheromones Do Not Cause Social Suppression of Reproduction in Male and Female Naked Mole-rats (*Heterocephalus glaber*)' in *Journal of Reproduction and Fertility* (1993), pp. 225–30

Chapter Eight: Primate politics

187 'Sputnik-era research grant to swell my pride': Alison Jolly, *Lords and Lemurs* (Houghton Mifflin, 2004), p. 3

188 males don't pose much of a threat to them: Christine M. Drea and Elizabeth S. Scordato, 'Olfactory Communication in the Ringtailed Lemur (*Lemur catta*): Form and Function of Multi-modal Signals' in *Chemical Signals in Vertebrates*, ed. by J. L. Hurst, R. J. Beynon, S. C. Roberts and T. Wyatt (2008), pp. 91–102

188 they keep sniffing the scent marks of the neighbours: Anne S. Mertl-Millhollen, 'Scent Marking as Resource Defense by Female *Lemur catta*' in *American Journal of Primatology*, 68: 6 (2006)

188 'exceptionally aggressive': Marie J. E. Charpentier and Christine M. Drea, 'Victims of Infanticide and Conspecific Bite Wounding in a Female-dominant Primate: A Long-term Study' in *PLoS One*, 8: 12 (2013), p. 5

189 Some males even die as a result of female violence: ibid., pp. 1–8

189 'put him in his place', 'where all females can be said to be dominant over all males': Alison Jolly, *Lemur Behaviour: A Madagascar Field Study* (University of Chicago Press, 1966), p. 155

189 'a particularly exciting glimpse of our history': ibid., p. 3

190 'the not-quite-monkey': ibid.

190 'depends ultimately primarily on the power of males': S. Washburn and D. Hamburg, 'Aggressive Behaviour in Old World Monkeys and Apes' in *Primates – Studies in Adaptation and Variability*, ed. by P. C. Jay (Holt, Rinehart and Winston, 1968)

190 'Female baboons are always dominated by their males': Vinciane Despret, 'Culture and Gender Do Not Dissolve into How Scientists "Read" Nature: Thelma Rowell's Heterodoxy' in *Rebels, Mavericks and Heretics in Biology*, ed. by Oren Harman and Michael R. Dietrich (Yale University Press, 2008)

191 'The search for our ancestry': Dale Peterson and Richard Wrangham, *Demonic Males: Apes and the Origins of Human Violence* (Mariner Books, 1997)

191 'the myth of the "typical" primate': Karen B. Strier, 'The Myth of the Typical Primate' in *American Journal of Physical Anthropology* (1994)

191 Their behaviour is actually highly derived: Anthony Di Fiore and Drew Rendall, 'Evolution of Social Organization: A Reappraisal for Primates by Using Phylogenetic Methods' in *PNAS*, 91: 21 (1994), pp. 9941–5

191n All but one of these genera are from the Old World: Karen B. Strier, 'New World Primates, New Frontiers: Insights from the Woolly Spider Monkey, or Muriqui (*Brachyteles arachnoides*)' in *International Journal of Primatology*, 11 (1990), pp. 7–19

192 'intellectually isolated': Rebecca J. Lewis, 'Female Power in Primates and the Phenomenon of Female Dominance' in *Annual Review of Anthropology*, 47 (2018), pp. 533–51

193 selection favours an intermediate body size and powerful long legs: Richard R. Lawler, Alison F. Richard and Margaret A. Riley, 'Intrasexual Selection in Verreaux's Sifaka (*Propithecus verreauxi verreauxi*)' in *Journal of Human Evolution*, 48 (2005), pp. 259–77

194 over five cubic centimetres: J. A. Parga, M. Maga and D. Overdorff, 'High-resolution X-ray Computed Tomography Scanning of Primate Copulatory Plugs' in *American Journal of Physical Anthropology*, 129: 4 (2006), pp. 567–76

194 no difference in size between the sexes: A. E. Dunham and V. H. W. Rudolf, 'Evolution of Sexual Size Monomorphism: The Influence of Passive Mate Guarding' in *Journal of Evolutionary Biology*, 22 (2009), pp. 1376–86

194 'Holy Grail' of lemur research: Amy E. Dunham, 'Battle of the Sexes: Cost Asymmetry Explains Female Dominance in Lemurs' in *Animal Behaviour*, 76 (2008), pp. 1435–9

194n 'semen becomes gelatinous': Alan F. Dixson and Matthew J. Anderson, 'Sexual Selection, Seminal Coagulation and Copulatory Plug Formation in Primates' in *Folia Primatologica*, 73 (2002), pp. 63–9

195 'masculinized' genitalia: Christine M. Drea, 'Endocrine Mediators of Masculinization in Female Mammals' in *Current Directions in Psychological Science*, 18: 4 (2009)

195 'skin identical in composition to that of the male's scrotum': Christine M. Drea, 'External Genital Morphology of the Ringtailed Lemur (*Lemur catta*): Females Are Naturally "Masculinized"', in *Journal of Morphology*, 269 (2008), pp. 451–63

196 monitored the subsequent level of rough and tumble play: Nicholas M. Grebe, Courtney Fitzpatrick, Katherine Sharrock, Anne Starling and Christine M. Drea, 'Organizational and Activational Androgens, Lemur Social Play, and the Ontogeny of Female Dominance' in *Hormones and Behavior* (Elsevier), 115 (2019)

196 up to 60 per cent of births are stillborn: S. E. Glickman, G. R. Cunha, C. M. Drea, A. J. Conley and N. J. Place, 'Mammalian Sexual Differentiation: Lessons from the Spotted Hyena' in *Trends in Endocrinology and Metabolism*, 17: 9 (2006), pp. 349–56

196 their aggression levels are so high: Charpentier and Drea, 'Victims of Infanticide and Conspecific Bite Wounding in a Female-dominant Primate'

197 some seventy-four million years ago: L. Pozzi, J. A. Hodgson, A. S. Burrell, K. N. Sterner, R. L. Raaum and T. R. Disotell, 'Primate Phylogenetic Relationships and Divergence Dates Inferred from Complete Mitochondrial Genomes' in *Molecular Phylogenetics and Evolution*, 75 (2014), pp. 165–83

198 'the boss': Frans de Waal, *Chimpanzee Politics: Power and Sex Among Apes* (Johns Hopkins University Press, 1982), p. 185

198 'There is great power in Mama's gaze': ibid., p.55

198n 'unconscious anthropomorphism': Thelma Rowell, 'The Concept of Social Dominance' in *Behavioural Biology* (June 1974), pp. 131–54

199 'feel small': Frans de Waal, *Mama's Last Hug* (Granta, 2019)

199 'respect from below rather than intimidation and strength from above': de Waal, *Chimpanzee Politics*

199n: She declined: Despret, 'Culture and Gender Do Not Dissolve into How Scientists "Read" Nature'

200 'a born diplomat': de Waal, *Mama's Last Hug*, p. 23

200 'formal hierarchy': ibid., p. 38

201 'Her wish was the colony's wish': ibid.

201 strongly influenced by the support of high-ranking females: Barbara Smuts, 'The Evolutionary Origins of Patriarchy' in *Human Nature*, 6 (1995), p. 9

201 Females will prevent certain males from joining the group: Smuts, 'The Evolutionary Origins of Patriarchy'

201 provides females with the authority to lead their group: Peter M. Kappeler, Claudia Fichtel, Mark van Vugt and Jennifer E. Smith, 'Female leadership: A Transdisciplinary Perspective' in *Evolutionary Anthropology* (2019), pp. 160–63

201 Challenging the age-old assumption: Jean-Baptiste Leca, Noëlle Gunst, Bernard Thierry and Odile Petit, 'Distributed Leadership in Semifree-ranging White-faced Capuchin Monkeys' in *Animal Behaviour*, 66 (Jan. 2003), pp. 1045–52

202 'Primate females seem biologically unprogrammed to dominate political systems': Lionel Tiger, 'The Possible Biological Origins of Sexual Discrimination' in *Biosocial Man*, ed. by D. Brothwell (Eugenics Society, London, 1970)

202 chipping away at the assumed autonomy of the alpha male: Jennifer E. Smith, Chelsea A. Ortiz, Madison T. Buhbe and Mark van Vugt, 'Obstacles and Opportunities for Female Leadership in Mammalian Societies: A Comparative Perspective' in *Leadership Quarterly*, 31: 2 (2020)

202–3 sex that stays in their birth group: Richard Wrangham, 'An Ecological Model of Female-bonded Primate Groups' in *Behaviour*, 75 (1980), pp. 262–300

203 'the most wretched and least independent of any non-human primate': Sarah Blaffer Hrdy, *The Woman That Never Evolved* (Harvard University Press, 1981), p. 101

203 'excessively paternalistic': ibid.

204 'a gift to the feminist movement': Frans de Waal, 'Bonobo Sex and Society' in *Scientific American* (1995), pp. 82–8

205 producing her first offspring: ibid.

208 'penis-fencing': ibid.

208 one out of three copulations in the wild: ibid.

209 The soliciting female will point backward with a foot: Pamela Heidi Douglas and Liza R. Moscovice, 'Pointing and Pantomime in Wild Apes? Female Bonobos Use Referential and Iconic Gestures to Request Genito-genital Rubbing' in *Scientific Reports*, 5 (2015)

212 ideologies of male dominance/female subordinance: Smuts, 'The Evolutionary Origins of Patriarchy'

212 'The roots of patriarchy lie in our pre-human past': ibid.

213 'male chivalry' or 'female feeding priority coupled with male social dominance': Frans de Waal and Amy R. Parish, 'The Other "Closest Living Relative": How Bonobos (*Pan paniscus*) Challenge Traditional Assumptions about Females, Dominance, Intra- and Intersexual Interactions, and Hominid Evolution' in *Annals of the New York Academy of Sciences* (2006)

213 'politically driven illusion engendered by feminism': ibid.

213 'imagine that we had never heard of chimpanzees': de Waal, 'Bonobo Sex and Society'

Chapter Nine: Matriarchs and menopause

216 more surface area for complex thought processes: Patrick R. Hof, Rebecca Chanis and Lori Marino, 'Cortical Complexity in Cetacean Brains' in *American Association for Anatomy*, 287A: 1 (Oct. 2005), pp. 1142–52

218 The four species of toothed whale to go through menopause: Samuel Ellis, Daniel W. Franks, Stuart Nattrass, Thomas E. Currie, Michael A. Cant, Deborah Giles, Kenneth C. Balcomb and Darren P. Croft, 'Analyses of Ovarian Activity Reveal Repeated Evolution of Post-reproductive Lifespans in Toothed Whale' in *Scientific Reports*, 8: 1 (2018)

219 'The realization slowly dawned': Howard Garrett, 'Orcas of the Salish Sea', Orca Network, http://www.orcanetwork.org [accessed Oct. 2019]

220 evolutionary upshot of the human male's preference for younger females: Richard A. Morton, Jonathan R. Stone and Rama S. Singh, 'Mate Choice and the Origin of Menopause' in *PLoS Computational Biology*, 9: 6 (2013)

220 Proposed in 1998: K. Hawkes et al., 'Grandmothering, Menopause and the Evolution of Human Life Histories' in *PNAS*, 95: 3 (1998), pp. 1336–9

222 phenomenal photographic memories: Marina Kachar, Ewa Sowosz and André Chwalibog, 'Orcas are Social Mammals' in *International Journal of Avian & Wildlife Biology*, 3: 4 (2018), pp. 291–5

223 playing elephant families audio recordings: Karen McComb, Cynthia Moss, Sarah M. Durant, Lucy Baker and Soila Sayialel, 'Matriarchs as Repositories of Social Knowledge in African Elephants' in *Science*, 292: 5516 (2001), pp. 491–4

224 ovaries are in fact still functioning in their seventies: F. J. Stansfield, J. O Nöthling and W. R. Allen, 'The Progression of Small-follicle Reserves in the Ovaries of Wild African Elephants (*Loxodonta africana*) from Puberty to Reproductive Senescence' in *Reproduction, Fertility and Development* (CSIRO publishing), 25: 8 (2013), pp. 1165–73

225 twelve-year study on food sharing: Brianna M. Wright, Eva M. Stredulinsky, Graeme M. Ellis and John K. B. Ford, 'Kin-directed Food Sharing Promotes Lifetime Natal Philopatry of Both Sexes in a Population of Fish-eating Killer Whales, *Orcinus orca*' in *Animal Behaviour*, 115 (2016), pp. 81–95

225 twice as likely to die: Darren P. Croft, Rufus A. Johnstone, Samuel Ellis, Stuart Nattrass, Daniel W. Franks, Lauren J. N. Brent, Sonia Mazzi, Kenneth C. Balcomb, John K. B. Ford and Michael A. Cant, 'Reproductive Conflict and the Evolution of Menopause in Killer Whales' in *Current Biology*, 27: 2 (2017), pp. 298–304

226 'reproductive conflict hypothesis': M. A. Cant, R. A. Johnstone and A. F. Russell, 'Reproductive Conflict and the Evolution of Menopause' in *Reproductive Skew in Vertebrates*, ed. by

R. Hager and C. B. Jones (Cambridge University Press, 2009), pp. 24–52

230 linked with the female orca's increased social and leadership skills: Bruno Cozzi, Sandro Mazzariol, Michela Podestà, Alessandro Zotti and Stefan Huggenberger, 'An Unparalleled Sexual Dimorphism of Sperm Whale Encephalization' in *International Journal of Comparative Psychology*, 29: 1 (2016)

230 Their cerebrum makes up 81.5 per cent of brain volume: Lori Marino, Naomi A. Rose, Ingrid Natasha Visser, Heather Rally, Hope Ferdowsian and Veronika Slootsky, 'The Harmful Effects of Captivity and Chronic Stress on the Well-being of Orcas (*Orcinus orca*)' in *Journal of Veterinary Behavior*, 35 (2020), pp. 69–82

230n Sperm whales show an even greater discrepancy of EQ between the sexes: ibid.

231 'socially complex brainiacs': Lori Marino, 'Dolphin and Whale Brains: More Evidence for Complexity', YouTube, https://www.youtube.com/watch?v=4SOzhyU3jMo

233 elevated levels of influence, knowledge and perception: Phyllis C. Lee and C. J. Moss, 'Wild Female African Elephants (*Loxodonta africana*) Exhibit Personality Traits of Leadership and Social Integration' in *Journal of Comparative Psychology*, 126: 3 (2012), pp. 224–32

Chapter Ten: Sisters are doing it for themselves

238 'the largest proportion of "homosexual animals" in the world': Jon Mooallem, 'Can Animals Be Gay?' in *New York Times*, 31 March 2010

239 their unconventional coupling went undetected until 2008: Lindsay C. Young, Brenda J. Zaun and Eric A. Vanderwurf, 'Successful Same-sex Pairing in Laysan Albatross' in *Biology Letters*, 4: 4 (2008), pp. 323–5

240 'promiscuity, polygamy and polyandry are unknown in this species': Mooallem, 'Can Animals Be Gay?'

242–3 'the greatest hockey player ever': Jack Falla, 'Wayne Gretzky' in *The Top 100 NHL Players of All Time*, ed. by Steve Dryden (McClelland and Stewart, 1998)

244 intimate physical gestures strengthen the birds' pair bond: Inna Schneiderman, Orna Zagoory-Sharon and Ruth Feldman, 'Oxytocin During the Initial Stages of Romantic Attachment: Relations to Couples' Interactive Reciprocity' in *Psychoneuroendocrinology*, 37: 8 (2012), pp. 1277–85

244 cooperative breeding and low divorce rates: Elspeth Kenny, Tim R. Birkhead and Jonathan P. Green, 'Allopreening in Birds is Associated with Parental Cooperation Over Offspring Care and Stable Pair Bonds Across Years' in *Behavioural Ecology* (ISBE, 2017), pp. 1142–8

245 are projected to have all but disappeared: J. D. Baker, C. L. Littman and D. W. Johnston, 'Potential Effects of Sea Level Rise on the Terrestrial Habitats of Endangered and Endemic Megafauna in the North-western Hawaiian Islands' in *Endangered Species Research*, 4 (2006), pp. 1–10

245 represented 14 per cent of the breeding pairs: George L. Hunt and Molly Warner Hunt, 'Female-Female Pairing in Western Gulls (*Larus occidentalis*) in Southern California' in *Science*, 196 (1977), pp. 1466–7

245 similar reports amongst roseate tern: Ian C. T. Nisbet and Jeremy J. Hatch, 'Consequences of a Female-biased Sex-ratio in a Socially Monogamous Bird: Female-female Pairs in the Roseate Tern *Sterna dougallii*' in *International Journal of Avian Science* (1999)

247 nano-scale fibrils: Hadi Izadi, Katherine M. E. Stewart and Alexander Penlidis, 'Role of Contact Electrification and Electrostatic Interactions in Gecko Adhesion' in *Journal of the Royal Society, Interface*, 11: 98 (2014)

247 to develop 'astronaut anchors': Elizabeth Landau, 'Gecko Grippers Moving On Up', NASA, 12 April 2015, https://www.nasa.gov/jpl/gecko-grippers-moving-on-up

248 female-only club of around one hundred known vertebrates: Kate L. Laskowski, Carolina Doran, David Bierbach, Jens Krause and Max Wolf, 'Naturally Clonal Vertebrates Are an Untapped Resource in Ecology and Evolution Research' in *Nature Ecology & Evolution* (2019), pp. 161–9

249 'queen of questions': Graham Bell, *The Masterpiece of Nature* (University of California Press, 1982)

249n potential for doubling production in an all-female species: Joan Roughgarden, *Evolution's Rainbow* (University of California Press, 2004), p. 17

251 'mutational meltdown': Logan Chipkin, Peter Olofsson, Ryan C. Daileda and Ricardo B. R. Azevedo, 'Muller's Ratchet in Asexual Populations Doomed to Extinction', eLife, 13 Nov. 2018, https://doi.org/10.1101/448563

251 'evolutionary dead ends' on the tree of life: Malin Ah-King, 'Queer Nature: Towards a Non-normative View on Biological Diversity' in *Body Claims*, ed. by J. Bromseth, L. Folkmarson Käll and K. Mattsson (Centre for Gender Research, Uppsala University, 2009)

251 'evolutionary scoundrels': J. Maynard Smith, *The Evolution of Sex* (Cambridge University Press, 1978)

252 DNA from more than five hundred different other species: C. Boschetti, A. Carr, A. Crisp, I. Eyres, Y. Wang-Koh, E. Lubzens, T. G. Barraclough, G. Micklem and A. Tunnacliffe, 'Biochemical Diversification through Foreign Gene Expression in Bdelloid Rotifers' in *PLoS Genetics* (2012)

253 in better evolutionary health than would be expected: Maurine Neiman, Stephanie Meirmans and Patrick G. Meirmans, 'What Can Asexual Lineage Age Tell Us about the Maintenance of Sex?' in *The Year in Evolutionary Biology* (2009), vol. 1168, issue 1, pp. 185–200

253 she incorporates bits of her used stored 'stolen' sperm: Robert D. Denton, Ariadna E. Morales and H. Lisle Gibbs, 'Genome-specific Histories of Divergence and Introgression Between an

Allopolyploid Unisexual Salamander Lineage and Two Ancestral Sexual Species' in *Evolution* (2018)

254 some variation, even in genetically identical lines: Laskowski,
Doran, Bierbach, Krause and Wolf, 'Naturally Clonal Vertebrates Are an Untapped Resource'

254 hybrid of two closely related species: V. Volobouev and G. Pas-
teur, 'Chromosomal Evidence for a Hybrid Origin of Diploid
Parthenogenetic Females from the Unisexual-bisexual *Lepidodactylus lugubris* Complex' in *Cytogenetics and Cell Genetics*,
63 (1993), pp. 194–9

254 helps limit the impact of inbreeding over time: Laskowski,
Doran, Bierbach, Krause and Wolf, 'Naturally Clonal Vertebrates are an Untapped Resource'

254 'homosexual behaviour': Yehudah L. Werner, 'Apparent Homo-
sexual Behaviour in an All-female Population of a Lizard,
Lepidodactylus lugubris and its Probable Interpretation' in
Zeitschrift für Tierpsychologie, 54 (1980), pp. 144–50

255 even adopting the so-called 'doughnut' sexual position: David
Crews, ' "Sexual" Behavior in Parthenogenetic Lizards (*Cnemidophorus*)' in *PNAS*, 77: 1 (1980), pp. 499–502

255 typically switched on or off in males by testosterone: L. A.
O'Connell, B. J. Matthews, D. Crews, 'Neuronal Nitric Oxide
Synthase as a Substrate for the Evolution of Pseudosexual Behav-
iour in a Parthenogenetic Whiptail Lizard' in *Journal of
Neuroendocrinology*, 23 (2011), pp. 244–53

255 switch sex roles and maximize their fertility: David Crews,
'The Problem with Gender' in *Psychobiology*, 16: 4 (1988), pp.
321–34

255 the all-female species behaved very differently: Beth E. Leuck,
'Comparative Burrow Use and Activity Patterns of Parthenogenetic and Bisexual Whiptail Lizards (Cnemidophorus: Teiidae)'
in *Copeia*, 2 (1982), pp. 416–24

256 revealed that evolution need not favour sexual reproduction:
Sarah P. Otto and Scott L. Nuismer, 'Species Interactions and the
Evolution of Sex' in *Science*, 304: 5673 (2004), pp. 1018–20

256 the first all-female termite society: T. Yashiro, N. Lo, K. Kobayashi, T. Nozaki, T. Fuchikawa, N. Mizumoto, Y. Namba and K. Matsuura, 'Loss of Males from Mixed-sex Societies in Termites' in *BMC Biology*, 16 (2018)

257 'anti-capitalist anarchists': Lisa Margonelli, *Underbug: An Obsessive Tale of Termites and Technology* (Scientific American, 2018)

257 by literally 'closing the sperm gates': Toshihisa Yashiro and Kenji Matsuura, 'Termite Queens Close the Sperm Gates of Eggs to Switch from Sexual to Asexual Reproduction' in *PNAS*, 111: 48 (2014), pp. 17212–17

258 'males are dispensable in advanced animal societies': Yashiro, Lo, Kobayashi, Nozaki, Fuchikawa, Mizumoto, Namba and Matsuura, 'Loss of Males from Mixed-sex Societies'

258 there was no 'DNA of male origin': Roger Highfield, 'Shark's Virgin Birth Stuns Scientists', *Telegraph*, 23 May 2007

259 alternative sexual strategy for ancient vertebrates: Warren Booth and Gordon W. Schuett, 'The Emerging Phylogenetic Pattern of Parthenogenesis in Snakes' in *Biological Journal of the Linnean Society* (2015), pp. 1–15

259 female sawfish had started to clone themselves: Andrew T. Fields, Kevin A. Feldheim, Gregg R. Poulakis and Demian D. Chapman, 'Facultative Parthenogenesis in a Critically Endangered Wild Vertebrate' in *Current Biology* (Cell Press), 25: 11 (2015), pp. 446–7

260 enough to get the same genetic advantages: Kat McGowan, 'When Pseudosex is Better Than the Real Thing', Nautilus, Nov. 2016, https://nautil.us/issue/42/fakes/when-pseudosex-is-better-than-the-real-thing

260 conservationists are seeing some recovery: Fields, Feldheim, Poulakis and Chapman, 'Facultative Parthenogenesis in a Critically Endangered Wild Vertebrate'

261n doubt was cast on the validity of his experiment: N. I. Werthessen, 'Pincogenesis – Parthenogenesis in Rabbits by Gregory Pincus' in *Perspectives in Biology and Medicine*, 18: 1 (1974), pp. 86–93

Chapter Eleven: Beyond the binary

263 'Where does he do his barnacles?': Jean Deutsch, 'Darwin and Barnacles' in *Comptes Rendus Biologies*, 333: 2 (2010), pp. 99–106

263 'the probosciform penis' is 'wonderfully developed', 'lies coiled up', 'when fully extended': Charles Darwin, *Living Cirripedia: A monograph of the sub-class Cirripedia, with figures of all the species. The Lepadidæ; or, pedunculated cirripedes* (Ray Society, 1851), pp. 231–2

264 the microscopic 'parasites' were the male: ibid., pp. 231–2

264 'mere bags of spermatozoa', 'half embedded': Charles Darwin, letter to J. S. Henslow, 1 April 1848, in *The Correspondence of Charles Darwin*, ed. by Frederick Burkhardt and Sydney Smith (Cambridge University Press, 1988), vol. 4, p. 128

264 'an hermaphrodite species must pass into': Charles Darwin, letter to Joseph Hooker, 10 May 1848, in *Charles Darwin's Letters: A Selection, 1825–1859*, ed. by Frederick Burkhardt (Cambridge University Press, 1998), p. xvii

265 'Truly the schemes & wonders of nature are illimitable': Charles Darwin, letter to Charles Lyell, 14 Sept. 1849, in *The Life and Letters of Charles Darwin*, ed. by Francis Darwin (D. Appleton & Co., 1896), vol. 1, p. 345

265 Darwin was indeed right: Hsiu-Chin Lin, Jens T. Høeg, Yoichi Yusa and Benny K. K. Chan, 'The Origins and Evolution of Dwarf Males and Habitat Use in Thoracican Barnacles' in *Molecular Phylogenetics and Evolution*, 91 (2015), pp. 1–11

266 'potential hermaphrodite' that 'emphasizes male function': Yoichi Yusa, Mayuko Takemura, Kota Sawada and Sachi Yamaguchi, 'Diverse, Continuous, and Plastic Sexual Systems in Barnacles' in *Integrative and Comparative Biology* 53: 4 (2016), pp. 701–12

266 their sexual expression is considered more of a continuum: ibid.

267 'The biggest error of biology today': Joan Roughgarden, *Evolution's Rainbow* (University of California Press, 2004), p. 17

267 'suppressing the full story of gender and sexuality': ibid., p. 128

267 'The true story of nature is profoundly empowering': ibid., p. 181

267 more overlap between the sexes than is commonly recognized: Malin Ah-King, 'Sex in an Evolutionary Perspective: Just Another Reaction Norm' in *Evolutionary Biology*, 37 (2010), pp. 234–46

267 homosexual activity is denigrated to an inconvenient 'error': Roughgarden, *Evolution's Rainbow*, p. 127

268 homosexual activity in over three hundred species of vertebrate: Bruce Bagemihl, *Biological Exuberance: Animal Homosexuality and Natural Diversity* (Stonewall Inn Editions, 2000)

268 'a social-inclusionary trait': Roughgarden, *Evolution's Rainbow*, p. 27

268 These family units can be either aggressive or harmonious: ibid., pp. 134–5

268 animal families frequently fail to conform to this 'Noah's Ark view': Bagemihl, *Biological Exuberance*

268n much like gang members 'swearing an oath': Volker Sommer and Paul L. Vasey (eds), *Homosexual Behaviour in Animals: An Evolutionary Perspective* (Cambridge University Press, 2006)

268n understood as a diverse collection of multicausal phenomena: Aldo Poiani, *Animal Homosexuality: A Biosocial Perspective* (Cambridge University Press, 2010)

268n 'a 2019 study postulated that': Julia D. Monk et al, 'An Alternative Hypothesis for the Evolution of Same-Sex Sexual Behaviour in Animals' in *Nature, Ecology and Evolution* 3 (2019), pp.1622–31

269 'the appearance, behaviour, and life history of a sexed body': Roughgarden, *Evolution's Rainbow*, p. 27

269 Many academics welcomed Roughgarden's radical proposal: Patricia Adair Gowaty, 'Sexual Natures: How Feminism Changed Evolutionary Biology' in *Signs* 28: 3, p. 901; and Ellen Ketterson, 'Do Animals Have Gender?' in *Bioscience* 55: 2 (2005), pp. 178–80

269 'is an elite male heterosexual narrative': Roughgarden, *Evolution's Rainbow*, p. 234

269 Her insistence that . . . sexual selection is 'false': ibid., p. 5

269 Darwin's 'conceptual rot': ibid., p. 181

270 an unpopular idea that prompted no less than forty biologists to write: Sarah Blaffer Hrdy, 'Sexual Diversity and the Gender Agenda' in *Nature* (2004), p. 19–20; and Patricia Adair Gowaty, 'Standing on Darwin's Shoulders: The Nature of Selection Hypotheses' in *Current Perspectives on Sexual Selection: What's Left After Darwin?*, ed. by Thierry Hoquet (Springer, 2015)

270 What conceptual butterfly finally emerges: Hoquet, *Current Perspectives on Sexual Selection*

270 reconsider the linear relationship between sex, sexuality and gender: Malin Ah-King, 'Queer Nature: Towards a Non-normative View on Biological Diversity' in *Body Claims*, ed. by Janne Bromseth, Lisa Folkmarson Käll and Katarina Mattsson (Centre for Gender Research, Uppsala University, 2009), pp. 227–8

273 replaced by ripe egg-filled ovaries: Logan D. Dodd, Ewelina Nowak, Dominica Lange, Coltan G. Parker, Ross DeAngelis, Jose A. Gonzalez and Justin S. Rhodes, 'Active Feminization of the Preoptic Area Occurs Independently of the Gonads in *Amphiprion ocellaris*' in *Hormones and Behavior*, 112 (2019), pp. 65–76

277 a spectrum of continuous variation in sexual characteristics: Mary Jane West-Eberhard, *Developmental Plasticity and Evolution* (Oxford University Press, 2003)

277 'sex-reversed': R. Jiménez, M. Burgos, L. Caballero and R. Diaz de la Guardia, 'Sex Reversal in a Wild Population of *Talpa occidentalis*' in *Genetics Research*, 52: 2 (Cambridge, 1988), pp. 135–40

277 'hermaphrodite': A. Sánchez, M. Bullejos, M. Burgos, C. Hera, C. Stamatopoulos, R. Diaz de la Guardia and R. Jiménez, 'Females of Four Mole Species of Genus *Talpa* (insectivora, mammalia) are True Hermaphrodites with Ovotestes' in *Molecular Reproduction and Development*, 44 (1996), pp. 289–94

277 'intersex' or even a mixture of several terms: Francisca M. Real, Stefan A. Haas, Paolo Franchini, Peiwen Ziong, Oleg Simakov, Heiner Kuhl, Robert Schöpflin, David Heller, M-Hossein Moeinzadeh, Verena Heinrich, Thomas Krannich, Annkatrin Bressin, Michaela F. Hartman, Stefan A. Wudy and Dina K. N. Dechmann, Alicia Hurtado, Francisco J. Barrionuevo, Magdalena Schindler, Izabela Harabula, Marco Osterwalder, Micahel Hiller, Lars Wittler, Axel Visel, Bernd Timmermann, Axel Meyer, Martin Vingron, Rafael Jimémez, Stefan Mundlos and Darío G. Lupiáñez, 'The Mole Genome Reveals Regulatory Rearrangements Associated with Adaptive Intersexuality' in *Science*, 370: 6513 (Oct. 2020), pp. 208–14

277 'It is important to remember that the diversity found in nature': Janet L. Leonard, *Transitions Between Sexual Systems* (Springer, 2018), p. 14

277 'quantitative gender': ibid., p. 15

277 'conditional sex expression': ibid., p. 12

277 slow to evolve beyond basic binary definitions of sex: Ah-King, 'Queer Nature: Towards a Non-normative View on Biological Diversity'

278 individual variation within a sex is greater than the average variation between the sexes: David Crews, 'The (bi)sexual brain' in *EMBO Reports* (2012), pp. 1–6

278 A recent study into nematode worms: Hannah N. Lawson, Leigh R. Wexler, Hayley K. Wnuk, Douglas S. Portman, 'Dynamic, Nonbinary Specification of Sexual State in the C. elegans Nervous System', *Current Biology* (2020)

278 'These findings indicate that': ScienceDaily, University of Rochester Medical Center (10 August 2020), www.sciencedaily.com/releases/2020/08/200810140949.htm

279 Rather than thinking of the sexes as wholly different biological entities: Agustín Fuentes, 'Searching for the "Roots" of Masculinity in Primates and the Human Evolutionary Past', *Current Anthropology* 62: S23, S13–S25 (2021)

Conclusion: A natural world without prejudice

281 'Objective knowledge' is an oxymoron: Patricia Adair Gowaty, *Feminism and Evolutionary Biology: Boundaries, Intersections and Frontiers* (Springer Science and Business Media, 1997)

281 'There is a mask of theory': William Whewell, *The Philosophy of the Inductive Sciences: Founded Upon Their History* (1847), p. 42

282 'undue prominence to such as have evolved in the male line': Antoinette Brown Blackwell, *The Sexes Throughout Nature* (1875), p. 20

282 'organic equilibrium in physiological and psychological equivalence of the sexes': ibid., p. 56; and Patricia Adair Gowaty, *Feminism and Evolutionary Biology: Boundaries, Intersections and Frontiers* (Springer Science and Business Media, 1997), p. 45

282 preserve of the 'rational sex': Antoinette Brown Blackwell, Darwin Correspondence Project (University of Cambridge) https://www.darwinproject.ac.uk/antoinette-brown-blackwell [accessed: April 2021]

282 'The road not taken': Sarah Blaffer Hrdy, *Mother Nature* (Ballantine, 1999), p. 22

282n Darwin's response to her was addressed 'Dear Sir . . . ': Brown Blackwell, Darwin Correspondence Project (University of Cambridge), https://www.darwinproject.ac.uk/antoinette-brown-blackwell [accessed: April 2021]

284 no reference to the published empirical criticism: Paula Vasconcelos, Ingrid Ahnesjö, Jaelle C. Brealey, Katerina P. Günter, Ivain Martinossi-Allibert, Jennifer Morinay, Mattias Siljestam and Josefine Stångberg, 'Considering Gender-biased Assumptions in Evolutionary Biology' in *Evolutionary Biology*, 47 (2020), pp. 1–5

284 'did not reflect the shift happening in the scientific community': Linda Fuselier, Perri K. Eason, J. Kasi Jackson and Sarah Spauldin, 'Images of Objective Knowledge Construction in

Sexual Selection Chapters of Evolution Textbooks' in *Science and Education*, 27 (2018), pp. 479–99

284 males act, whereas females react: Kristina Karlsson Green and Josefin A. Madjidian, 'Active Males, Reactive Females: Stereotypic Sex Roles in Sexual Conflict Research?' in *Animal Behaviour*, 81 (2011), pp. 901–7

284 still bandied about in academic literature: Brealey, Günter, Martinossi-Allibert, Morinay, Siljestam, Stångberg and Vasconcelos, 'Considering Gender-Biased Assumptions in Evolutionary Biology'

284 The 'type specimens' that define a species: Natalie Cooper, Alexander L. Bond, Joshua L. Davis, Roberto Portela Miguez, Louise Tomsett and Kristofer M. Helgen, 'Sex Biases in Bird and Mammal Natural History Collections' in *Proceedings of the Royal Society B*, 286 (2019)

284 many researchers avoid using females: Annaliese K. Beery and Irving Zucker, 'Males Still Dominate Animal Studies' in *Nature*, 465: 690 (2010)

284 males are considered more pure: Rebecca M. Shansky, 'Are Hormones a "Female Problem" for Animal Research?' in *Science*, 364: 6443, pp. 825–6

285 Fruit flies . . . still dominate sexual selection research: Marlene Zuk, Francisco Garcia-Gonzalez, Marie Elisabeth Herberstein and Leigh W. Simmons, 'Model Systems, Taxonomic Bias, and Sexual Selection: Beyond *Drosophila*' in *Annual Review of Entomology* (2014), pp. 321–38

285 hasn't stopped their eccentric kinks being extrapolated: ibid.

285 cautioned against the use of improper model systems: ibid.

286 'taxonomic chauvinism': ibid.

286 LGBTQ people in STEM are still statistically underrepresented: Jonathan B. Freeman, 'Measuring and Resolving LGBTQ Disparities in STEM' in *Policy Insights from the Behavioral and Brain Sciences* (2020), pp. 141–8

286 The stale old propaganda concerning gender disparities: Ben A. Barres, 'Does Gender Matter?' in *Nature* (2006), pp. 133–6

287 Sample sizes were often too small: Yao-Hua Law, 'Replication Failures Highlight Biases in Ecology and Evolution Science', *The Scientist*, 1 Aug. 2018, https://www.the-scientist.com/features/replication-failures-highlight-biases-in-ecology-and-evolution-science-64475

287 cherry picking statistically significant results: Hannah Fraser, Tim Parker, Shinichi Nakagawa, Ashley Barnett and Fiona Fidler, 'Questionable Research Practices in Ecology and Evolution' in *PLoS One*, 13: 7 (2018), pp. 1–16.

287 Funding bodies and editors of scientific publications: Hannah Fraser, Ashley Barnett, Timothy H. Parker and Fiona Fidler, 'The Role of Replication Studies in Ecology' in *Academic Practice in Ecology and Evolution* (2020), pp. 5197–206

288 'Biology is politics by any other means': Anne Fausto-Sterling, *Sexing the Body* (Basic Books, 2000)

SELECTED BIBLIOGRAPHY

Altmann, Jeanne, *Baboon Mothers and Infants* (Harvard University Press, 1980)

Arnqvist, Göran and Locke Rowe, *Sexual Conflict* (Princeton University Press, 2005)

Bagemihl, Bruce, *Biological Exuberance: Animal Homosexuality and Natural Diversity* (Stonewall Inn Editions, 2000)

Barlow, Nora (ed.), *The Autobiography of Charles Darwin 1809–1882* (Collins, 1958)

Birkhead, Tim, *Promiscuity: An Evolutionary History of Sperm Competition and Sexual Conflict* (Faber & Faber, 2000)

Blackwell, Antoinette Brown, *The Sexes Throughout Nature* (Putnam and Sons, 1875)

Bleier, Ruth (ed.), *Feminist Approaches to Science* (Pergamon Press, 1986)

Campbell, Bernard (ed.), *Sexual Selection and the Descent of Man 1871–1971* (Aldine-Atherton, 1972)

Choe, Jae, *Encyclopedia of Animal Behavior,* second edition (Elsevier, 2019)

Clutton-Brock, Tim, *Mammal Societies* (John Wiley and Sons, 2016)

Cronin, Helena, *The Ant and the Peacock* (Cambridge University Press, 1991)

Darwin, Charles, *Living Cirripedia: A monograph of the subclass Cirripedia, with figures of all the species. The Lepadidæ; or, pedunculated cirripedes* (Ray Society, 1851)

Darwin, Charles, *On the Origin of Species by Means of Natural Selection* (John Murray, 1859; Mentor Books, 1958)

Darwin, Charles, *The Descent of Man, and Selection in Relation to Sex* (John Murray, 1871; second edition 1979; Penguin Classics 2004)

Davies, N. B., *Dunnock Behaviour and Social Evolution* (Oxford University Press, 1992)

Dawkins, Richard, *The Selfish Gene* (Oxford University Press, 1976; new edition 1989)

Denworth, Lydia, *Friendship: The Evolution, Biology and Extra-ordinary Power of Life's Fundamental Bond* (Bloomsbury, 2020)

DeSilva, Jeremy (ed.), *A Most Interesting Problem: What Darwin's* Descent of Man *Got Right and Wrong about Human Evolution* (Princeton University Press, 2021)

de Waal, Frans, *Chimpanzee Politics: Power and Sex among Apes* (Johns Hopkins University Press, 1982)

de Waal, Frans, *Bonobo: The Forgotten Ape* (University of California Press, 1997)

de Waal, Frans, *The Bonobo and the Atheist: In Search of Humanism among the Primates* (W. W. Norton & Co., 2013)

de Waal, Frans, *Mama's Last Hug* (Granta, 2019)

Dixson, Alan F., *Primate Sexuality: Comparative Studies of the Prosimians, Monkeys, Apes, and Humans* (Oxford University Press, 2012)

Drickamer, Lee and Donald Dewsbury (eds), *Leaders in Animal Behaviour* (Cambridge University Press, 2010)

Eberhard, William G., *Sexual Selection and Animal Genitalia* (Harvard University Press, 1985)

Eberhard, William G., *Female Control: Sexual Selection by Cryptic Female Choice* (Princeton University Press, 1996)

Elgar, M. A. and J. M. Schneider, 'The Evolutionary Significance of Sexual Cannibalism' in Peter Slater et al. (eds), *Advances in the Study of Behavior,* volume 34 (Academic Press, 2004)

Fausto-Sterling, Anne, *Sexing the Body: Gender Politics and the Construction of Sexuality* (Basic Books, 2000)

Fedigan, Linda Marie, *Primate Paradigms: Sex Roles and Social Bonds* (University of Chicago Press, 1982)

Fine, Cordelia, *Testosterone Rex* (W. W. Norton & Co., 2017)

Fisher, Maryanne L., Justin R. Garcia and Rosemarie Sokol Chang (eds), *Evolution's Empress: Darwinian Perspectives on the Nature of Women* (Oxford University Press, 2013)

Fuentes, Agustin, *Race, Monogamy and Other Lies They Told You: Busting Myths about Human Nature* (University of California Press, 2012)

Gould, Stephen Jay, *The Flamingo's Smile: Reflections in Natural History* (W. W. Norton & Co., 1985)

Gowaty, Patricia (ed.), *Feminism and Evolutionary Biology: Boundaries, Intersections and Frontiers* (Springer, 1997)

Haraway, Donna J., *Primate Visions: Gender, Race, and Nature in the World of Modern Science* (Routledge, 1989)

Hayssen, Virginia and Teri J. Orr, *Reproduction in Mammals: The Female Perspective* (Johns Hopkins University Press, 2017)

Hoquet, Thierry (ed.), *Current Perspectives on Sexual Selection: What's Left After Darwin?* (Springer, 2015)

Hrdy, Sarah Blaffer, *The Langurs of Abu: Female and Male Strategies of Reproduction* (Harvard University Press, 1980)

Hrdy, Sarah Blaffer, *The Woman That Never Evolved* (Harvard University Press, 1981; second edition, 1999)

Hrdy, Sarah Blaffer, *Mother Nature: Maternal Instincts and How They Shape the Human Species* (Ballantine Books, 1999)

Hrdy, Sarah Blaffer, *Mothers and Others: The Evolutionary Origins of Mutual Understanding* (Harvard University Press, 2009)

Jolly, Alison, *Lemur Behaviour: A Madagascar Field Study* (University of Chicago Press, 1966)

Jolly, Alison, *Lords and Lemurs: Mad Scientists, Kings with Spears, and the Survival of Diversity in Madagascar* (Houghton Mifflin Company, 2004)

Kaiser, David and W. Patrick McCray (eds), *Groovy Science: Knowledge, Innovation, and American Counterculture* (University of Chicago Press, 2016)

Lancaster, Roger, *The Trouble with Nature: Sex in Science and Popular Culture* (University of California Press, 2003)

Leonard, Janet (ed.), *Transitions Between Sexual Systems: Understanding the Mechanisms of, and Pathways Between, Dioecy, Hermaphroditism and Other Sexual Systems* (Springer, 2018)

Margonelli, Lisa, *Underbug: An Obsessive Tale of Termites and Technology* (Scientific American, 2018)

Marzluff, John and Russell Balda, *The Pinyon Jay: Behavioral Ecology of a Colonial and Cooperative Corvid* (T. and A. D. Poyser, 1992)

Maynard Smith, John, *The Evolution of Sex* (Cambridge University Press, 1978)

Milam, Erika Lorraine, *Looking for a Few Good Males: Female Choice in Evolutionary Biology* (Johns Hopkins University Press, 2010)

Morris, Desmond, *The Naked Ape* (Jonathan Cape, 1967)

Mundy, Rachel, *Animal Musicalities: Birds, Beasts, and Evolutionary Listening* (Wesleyan University Press, 2018)

Oldroyd, D. R. and K. Langham (eds), *The Wider Domain of Evolutionary Thought* (D. Reidel Publishing Company, 1983)

Poiani, Aldo, *Animal Homosexuality: A Biosocial Perspective* (Cambridge University Press, 2010)

Prum, Richard O., *The Evolution of Beauty: How Darwin's Forgotten Theory of Mate Choice Shapes the Animal World Around Us* (Anchor Books, 2017)

Rees, Amanda, *The Infanticide Controversy: Primatology and the Art of Field Science* (University of Chicago Press, 2009)

Rice, W. and S. Gavrilets (eds), *The Genetics and Biology of Sexual Conflict* (Cold Spring Harbor Laboratory Press, 2015)

Rosenthal, Gil G., *Mate Choice: The Evolution of Sexual Decision Making from Microbes to Humans* (Princeton University Press, 2017)

Roughgarden, Joan, *Evolution's Rainbow: Diversity, Gender, and Sexuality in Nature and People* (University of California Press, 2004)

Russett, Cynthia, *Sexual Science: The Victorian Construction of Womanhood* (Harvard University Press, 1991)

Ryan, Michael J., *A Taste for the Beautiful: The Evolution of Attraction* (Princeton University Press, 2018)

Saini, Angela, *Inferior: How Science Got Women Wrong – and the New Research That's Rewriting the Story* (Fourth Estate, 2017)

Schilthuizen, Menno, *Nature's Nether Regions: What the Sex Lives of Bugs, Birds and Beasts Tell Us About Evolution, Biodiversity and Ourselves* (Viking, 2014)

Schutt, Bill, *Eat Me: A Natural and Unnatural History of Cannibalism* (Profile Books, 2017)

Smuts, Barbara B., *Sex and Friendship in Baboons* (Aldine Publishing Co., 1986)

Sommer, Volker and Paul F. Vasey (eds), *Homosexual Behaviour in Animals: An Evolutionary Perspective* (Cambridge University Press, 2004)

Symons, Donald, *The Evolution of Human Sexuality* (Oxford University Press, 1979)

Travis, Cheryl Brown (ed.), *Evolution, Gender, and Rape* (MIT Press, 2003)

Travis, Cheryl Brown and Jacquelyn W. White (eds), *APA Handbook of the Psychology of Women: History, Theory, and Battlegrounds* (American Psychological Association, 2018)

Tutin, Caroline, *Sexual Behaviour and Mating Patterns in a Community of Wild Chimpanzees* (University of Edinburgh, 1975)

Viloria, Hilda and Maria Nieto, *The Spectrum of Sex: The Science of Male, Female and Intersex* (Jessica Kingsley Publishers, 2020)

Wallace, Alfred Russel, *Darwinism: An Exposition of the Theory of Natural Selection with Some of its Applications* (Macmillan & Co., 1889)

Wasser, Samuel K., *Social Behaviour of Female Vertebrates* (Academic Press, 1983)

West-Eberhard, Mary Jane, *Developmental Plasticity and Evolution* (Oxford University Press, 2003)

Whewell, William, *The Philosophy of the Inductive Sciences: Founded Upon Their History* (J. W. Parker, 1847)

Willingham, Emily, *Phallacy: Life Lessons from the Animal Penis* (Avery, 2020)

Wilson, E. O., *Sociobiology: The New Synthesis* (Harvard University Press, 1975; twenty-fifth-anniversary edition 2000)

Wrangham, Richard and Dale Peterson, *Demonic Males* (Houghton Mifflin, 1996)

Yamagiwa, Juichi and Leszek Karczmarski, *Primates and Cetaceans: Field Research and Conservation of Complex Mammalian Societies* (Springer, 2014)

Zuk, Marlene, *Sexual Selections: What We Can and Can't Learn about Sex from Animals* (University of California Press, 2002)

Zuk, Marlene and Leigh W. Simmons, *Sexual Selection: A Very Short Introduction* (Oxford University Press, 2018)

INDEX

baboons – *cont.*
parenting 137; savannah baboons
61, 67, 133–9, 141–2, 189; sex-
manipulation 139, 140; social class
135–9, 141, 166, 167
baculum 100
Baden, Dr Andrea 151–4
Bagemihl, Bruce 267–8
Balcolm, Ken 218
Balda, Russell xvi, xvii
baobab 186, 187, 195
barbary macaques: copulations 61;
paternity confusion 67
barklice 3; *Afrotogla* 3n; *Neotrogla* 3n
barnacles: continuum of sex 266;
Darwin's interest in 262–5, 277;
ovaries 265–6; penis 263
Bateman, Angus John 48–51; Bateman's
paradigm 48–51, 56, 59, 65, 69,
70–6, 284, 287
Bateman's gradient 49, 51
bats: fruit bats 125, 125n; genital
diversity 100; nurseries 150–1; penis
105; size dimorphism of *Ametrida
centurio* 6n; testes size 69
baubellum 118
bdelloid rotifer 251–3, 261; survival of
252–3
bearded dragons (*Pogona vitticeps*)
21–2
bed bugs (*Cimicidae*) 107–8, 107n
bees x, 164, 174, 257; all-female societies
256; arrhenotoky 260–1n
beluga whales (*Delphinapterus leucas*)
218n
Bempton Cliffs, Yorkshire 54–5, 59
bereavement, elephants and 234
bird-eating spiders 79–81
birds xviii–xix; allomothers 151;
allopreening 244; birdsong 160–3,
164; brain size 160; caring for
young 126; cloaca 56, 105, 106;
courtship rituals 28–32, 35–43;
forced copulations 57n, 106–7,
107n, 108, 110; genes 16, 18;
gynandromorphs 22–4;
homosexuality 238–46; mating 35,

37, 43, 46, 42–4, 104–6, 108–10,
206, 208–9; missing penises 105,
110–11; monogamy 53, 54, 58, 111,
238, 244; Neoaves 105, 110; penises
57n, 104, 105–7, 108–10; polyandry
53, 53n, 55–9, 66; sex-manipulation
139; sexual dimorphism 24; social
status 166; songbirds 24, 52, 53–4,
55–9, 160–3; *see also individual
species*
birds of paradise 35
Birkhead, Tim 54–5, 57, 71, 73;
Promiscuity 59
birth, oxytocin and 145
bisexuality 264, 276, 276n, 278; bonobo
208; and the brain 130–1, 149–50,
271–2, 273, 274, 276, 278;
hermaphrodites 276; Laysan
albatross 20, 238–45, 278; whiptail
lizard 254–5, 276
black-and-white ruffed lemur (*Varecia
variegate*) 151–4, 182; allomothers
151, 153–4
black-capped chickadees 57
black-headed owl monkey (*Aotus
nigriceps*) 122–5; nipples 125
black widow spider 83
blackbird, red-winged (*Agelaius
phoeniceus*) 52–4
Blackwell, Antoinette Brown 282n; *The
Sexes Throughout Nature* 282
blastocysts 141
blue tit 58
blue whale 6n
bluehead wrasse (*Thalassoma
bifasciatum*) 269, 270–1
bonding: female to female 211; mother-
baby 147–9
Bonnet, Charles 250–1n
bonobo (*Pan paniscus*) x, 204–14,
287–8; aggression 207–8; body
proportions 210; in captivity 213;
clitoris 205, 118; connection with
humans 210–11; female coalitions
206–7; G-G rubbing 205, 208, 209;
oral sex 208; orgasms 205;
penis-fencing 208; sex as social glue

Clarke, Dave 78–9, 81, 82, 85–7
climate change 260
clitoris 61–2, 116–19; action of
    androgens on 7, 9; bonobo 118,
    205; 'discovery of' 117; diversity
    61–2, 116, 118, 195; dolphin 116,
    118; first detailed anatomy of the
    118; fossa 4–5; homologue of penis
    64; human 118; lemur 195; moles 2;
    orgasms in primates 64; primates
    64, 118; 'pseudo-penis' 3, 4, 5, 7, 9,
    61, 195; spotted hyena 5, 7, 9, 61,
    195, 196
cloaca 56, 105, 106; cloacal kiss 105;
    cloacal winking 110
cloning x, 248–61, 276; desert grassland
    whiptails 255–6, 276; Dolly the
    sheep 249n; first creatures 26;
    mourning gecko 248–9, 254;
    sawfish 259, 260; sharks 258–60;
    termites 257–8
clownfish (anemonefish) 271, 272–6,
    278, 288
Clutton-Brock, Professor Tim 169,
    170–1, 172–3
cognition, evolution of 41, 155
colobus monkey 203
communication: Laysan albatross
    243–4; lion 44–5; seismic
    communication 90–4, 90n
confirmation bias xv, xvii, 72, 73
Coolidge, Harold J. 209–10
cooperative breeding 151; insects 174–5;
    meerkats 168, 172; naked mole rats
    177–80; termites 174–5
copulation: birds 35, 37, 43, 46, 42–4,
    104–6, 108–10, 206, 208–9; bonobo
    206, 208–9; chimpanzees 208, 209,
    213; dolphins 114–15, 116; ducks
    105–7, 108–10; forced copulations
    57n, 106–7, 106–7n, 108, 110, 114;
    fruit flies 48–50, 70–3, 74; golden
    orb weaver spider 77–9, 82–3, 98;
    greater sage grouse 35, 37, 43; killer
    wales 219n, 221, 225; mate-choice
    28–43; non-conceptive as male
    manipulation 64–7; praying mantis

97–8; red-winged blackbirds 52–4;
    remote copulation 84; sexual
    cannibalism 77–98; spiders 77–97;
    threesomes 83–4, 268; túngara frog
    43; *see also* masturbation; oral sex
copulatory plugs 84, 194, 194n
Cornell Lab of Ornithology 36
courtship: ducks 106, 109; greater sage
    grouse 28–32, 35–8, 40–1, 42–3;
    Laysan albatross 239, 278; lizards
    255; peacock spider 87–94; satin
    bowerbird 38–40, 41, 42; sexual
    pleasure 116, 118–19; spiders 83,
    87–94; superb fairy wren 55
coypu 140
Crews, David xx, 24–7, 276, 276n,
    277–8, 285
crickets: bush cricket 69n, 118–19;
    Mormon cricket 70; Texan field
    cricket 69n; semen 69n; sex 70; sex
    roles 75
Croft, Darren 219–20, 221, 222–3, 224,
    225, 232
cryptic female choice 112–13, 115, 120

damselfly 112; penis 102
Dantzker, Marc 36
dark fishing spider (*Dolomedes
    tenebrosus*) 95, 97
Darwin, Charles xi–xiv, xvi, xviii, xix,
    xx, 43, 72, 277–8; active males vs
    passive females xiii, 8; barnacles
    262–5, 277; cultural bias xii–xiii,
    xiv-xv, xviii–xix, xx, 33–4, 47, 55;
    *The Descent of Man, and Selection
    in Relation to Sex* xii, xv, 32, 46–7,
    78, 101, 102, 265; female choice
    32–3; genitalia 101; influence of
    Thomas Malthus 40, 40n;
    intrasexual competition 159–60;
    'law of battle' 156–7; male hunters
    154; male intellectual superiority
    230; maternal instinct 122, 124,
    142, 155; natural selection xi, xii,
    xv, 32, 40, 40n, 264; *On the Origin
    of Species* 160, 263; reproduction
    as harmonious affair 82; secondary

Lucy Cooke is an award-winning broadcaster and documentary filmmaker with a Masters in Zoology from the University of Oxford, where she was tutored by Richard Dawkins. She has presented prime-time series for BBC, ITV and National Geographic and is a regular on Radio 4, hosting her own *Power of . . .* series along with regular spots on *The Infinite Monkey Cage* and *Sue Perkins: Nature Table*. Lucy has written for the *Sunday Times, Telegraph, Mail on Sunday, New York Times* and *Wall Street Journal*. She is the author of two previous books, *A Little Book of Sloth*, which was a *New York Times* bestseller, and *The Unexpected Truth about Animals*, which was shortlisted for the Royal Society Insight Investment Science Book Prize and has been translated into seventeen languages.